T0214788

Canadian Mathematical Society
Société mathématique du Canada

More information about this series at http://www.springer.com/series/4318

Jennifer Hyndman • J. B. Nation

The Lattice of Subquasivarieties of a Locally Finite Quasivariety

 Springer

Jennifer Hyndman
Department of Mathematics and Statistics
University of Northern British Columbia
Prince George, BC, Canada

J. B. Nation
Department of Mathematics
University of Hawaii
Honolulu, HI, USA

ISSN 1613-5237 ISSN 2197-4152 (electronic)
CMS Books in Mathematics
ISBN 978-3-030-08651-0 ISBN 978-3-319-78235-5 (eBook)
https://doi.org/10.1007/978-3-319-78235-5

Printed on acid-free paper

This Springer imprint is published by the registered company Springer Nature Switzerland AG.
The registered company address is: Gewerbestrasse 11, 6330 Cham, Switzerland

In memory of Bjarni Jónsson

Preface

The book *Roads to Quoz*, by William Least Heat-Moon, describes that author's search for the mysterious reward that awaits the curious traveler, as much in the journey as the destination. This monograph describes the current authors' search, perhaps equally quixotic, to understand the mysteries of quasivarieties. As with Least Heat-Moon, we started with less ambitious objectives, but followed where the road led. You are invited to travel with us and share the pleasures.

This being a mathematical journey, we begin with some definitions.

A *variety* is the class of all algebras (of the same similarity type) that satisfy some given set of equations. For example, the variety of abelian groups is defined by the group axioms and commutativity:

$$x(yz) \approx (xy)z$$
$$x1 \approx x$$
$$xx^{-1} \approx 1$$
$$xy \approx yx.$$

Varieties, also known as *equational classes*, are characterized by being closed under homomorphic images, subalgebras, and direct products.

A *quasivariety* is the set of all algebras (of the same type) that satisfy some given set of implications

$$s_1 \approx t_1 \ \& \ \cdots \ \& \ s_{n-1} \approx t_{n-1} \to s_n \approx t_n$$

where each s_i, t_i is a term. We allow $n = 1$, so that equations count as implications, and varieties are also quasivarieties. For example, the quasivariety of all abelian groups with no element of order 5 is defined by the above equations and the implication

$$x^5 \approx 1 \to x \approx 1.$$

Quasivarieties are characterized by being closed under subalgebras, direct products, and ultraproducts.

Quasivarieties came into being as solutions to concrete algebra problems, perhaps beginning with Mal'cev's 1939 characterization of semigroups that can be embedded into groups [122, 123]. By the 1970s, quasivarieties of various algebraic systems were being studied in their own right. Of course, when one is studying quasivarieties of a specific type of algebra, say groups or lattices, then the structural properties of those algebras play a crucial role. But there are also some general methods that apply, and our goal in this monograph is to formalize and investigate some of those principles.

Both Garrett Birkhoff and A. I. Mal'cev suggested that quasivarieties could be studied in terms of their lattice of subquasivarieties [33, 125]. That is the approach we take, though as always, in the examples we must delve into the structure of particular types of algebras.

In this monograph, we restrict our attention to locally finite quasivarieties of finite type, i.e., with only finitely many operations and relations. This is in part because this investigation arose from our study of certain finitely generated quasivarieties and our desire to find a common method for dealing with those various situations. But also, the remarkable properties of locally finite quasivarieties seem to justify this restriction. Meanwhile, we have continued in parallel our investigation into subquasivariety lattices for general quasivarieties [17].

We can summarize the results as follows. Our goal is to develop methods for working with locally finite quasivarieties of finite type. These tools are used to analyze the structure of the lattice $L_q(\mathcal{K})$ of subquasivarieties of such a quasivariety $\mathcal{K}$. It turns out that the lattice $L_q(\mathcal{K})$ is both algebraic and dually algebraic, join semidistributive, and fermentable. Quasivarieties that are completely join irreducible or completely meet irreducible in $L_q(\mathcal{K})$ are characterized, and likewise quasivarieties that are completely join prime or completely meet prime.

For any finite algebra $\mathbf{T} \in \mathcal{K}$, there is a finite set $\mathcal{E}(\mathbf{T})$ of quasi-equations such that, for any subquasivariety $\mathcal{Q} \leq \mathcal{K}$, it is the case that $\mathbf{T} \notin \mathcal{Q}$ if and only if $\mathcal{Q}$ satisfies ε for some $\varepsilon \in \mathcal{E}(\mathbf{T})$. Moreover, given $\mathbf{T}$ and its congruence lattice Con $\mathbf{T}$, it is straightforward to find the quasi-equations in $\mathcal{E}(\mathbf{T})$. For each quasi-equation ε in $\mathcal{E}(\mathbf{T})$, there is a finite list $\mathbf{U}_1, \ldots, \mathbf{U}_k$ of finite algebras in $\mathcal{K}$ such that an algebra $\mathbf{S} \in \mathcal{K}$ satisfies ε if and only if $\mathbf{S}$ contains no $\mathbf{U}_i$ as a subalgebra.

Since $L_q(\mathcal{K})$ is algebraic, every subquasivariety of $\mathcal{K}$ is a meet of completely meet irreducible quasivarieties. For every completely meet irreducible quasivariety $\mathcal{M}$ in $L_q(\mathcal{K})$, there is a finite algebra $\mathbf{T}$, not in $\mathcal{M}$, such that

$\mathcal{M} = \langle \varepsilon \rangle$ for a quasi-equation ε in $\mathcal{E}(\mathbf{T})$. Quasivarieties $\mathcal{Q}$ that are finitely based relative to $\mathcal{K}$ are then the intersection of finitely many such quasivarieties $\langle \varepsilon_i \rangle$, with each ε_i in $\mathcal{E}(\mathbf{T}_i)$ for some $\mathbf{T}_i$ not in $\mathcal{Q}$. Thus, finitely based quasivarieties can also be characterized in terms of omitting finitely many subalgebras.

The methods are illustrated by applying them to quasivarieties of 1-unary and 2-unary algebras, lattices, abelian groups, and pure unary relational structures.

Prince George, BC, Canada   Jennifer Hyndman
Honolulu, HI, USA   J. B. Nation

Contents

List of Figures

List of Tables

CHAPTER 1

Introduction and Background

1.1. Introduction

The properties of a quasivariety are reflected in the structure of its lattice of subquasivarieties. For example, a subquasivariety $\mathcal{S}$ of a quasivariety $\mathcal{K}$ is finitely based relative to $\mathcal{K}$ if and only if $\mathcal{S}$ is dually compact in the lattice of subquasivarieties of $\mathcal{K}$. In order to understand how quasivarieties work, we need general methods to analyze their lattices of subquasivarieties.

Let $L_q(\mathcal{K})$ denote the lattice of subquasivarieties of a quasivariety $\mathcal{K}$. The goal of this monograph is to present algorithms that enable us to describe the structure of $L_q(\mathcal{K})$ when $\mathcal{K}$ is a quasivariety of locally finite structures of finite type. Most of our examples are unary algebras and pure unary relational structures, but we also consider quasivarieties of lattices and abelian groups. The theory developed here applies to structures of arbitrary finite type.

The basic properties of lattices $L_q(\mathcal{Q})$, without the locally finite and finite type restriction on $\mathcal{Q}$, were studied by Viktor Gorbunov and his Siberian school. These results are presented in Gorbunov's book [77]. Further refinements were obtained by Adaricheva and Nation [19]; see also [18]. The basic properties that are known to hold in general lattices of quasivarieties can be summarized as follows. (A more complete discussion is in the Appendix.) For any quasivariety $\mathcal{Q}$, the lattice $L_q(\mathcal{Q})$ of subquasivarieties

- is dually algebraic,
- is join semidistributive,
- is atomic lattice,
- supports an equaclosure operator,
- has the Jónsson-Kiefer property.

These properties can be derived most easily from representations of the subquasivariety lattices. Recall that a subset of a complete lattice is an *algebraic subset* if it is closed under arbitrary meets and nonempty directed joins. The characterization theorem of Gorbunov and Tumanov [79] says

J. Hyndman, J. B. Nation, *The Lattice of Subquasivarieties of a Locally Finite Quasivariety*, CMS Books in Mathematics, https://doi.org/10.1007/978-3-319-78235-5_1

that $L_q(\mathcal{Q})$ is isomorphic to the lattice $S_p(\mathbf{L}, R)$ of all algebraic subsets of an algebraic lattice $\mathbf{L}$ that are closed with respect to a distributive binary relation R; see Section 5.2 of Gorbunov [77].

The representation theorem of Adaricheva and Nation [19] says that $L_q(\mathcal{Q})$ is dually isomorphic to the congruence lattice of a semilattice with operators. Applying Hyndman *et al.* [87], we see that for any quasivariety $\mathcal{Q}$ there exist an algebraic lattice $\mathbf{L}$ and a monoid H of operators on $\mathbf{L}$, with each operator in H preserving arbitrary meets and nonempty directed joins, such that $L_q(\mathcal{Q})$ is isomorphic to $S_p(\mathbf{L}, H)$, the lattice of all H-closed algebraic subsets of $\mathbf{L}$.

It has been known for some time, however, that when $\mathcal{K}$ is a locally finite quasivariety of finite type, then there are additional restrictions. For varieties, we refer to Hobby and McKenzie [83] or Kiss and Kearnes [104]. With respect to quasivarieties, see Gorbunov [77] or Freese *et al.* [70]. For example, if $\mathcal{K}$ is locally finite and has finite type, and $L_q(\mathcal{K})$ is finite, then it is a lower bounded homomorphic image of a free lattice. This is not true for general quasivarieties: $L_q(\mathcal{Q})$ can be a finite lattice that is not lower bounded [19]. Also, for $\mathcal{K}$ locally finite of finite type, the lattices $L_q(\mathcal{K})$ satisfy quasi-equations that do not hold in general subquasivariety lattices [70, 77]. Indeed, a quasi-identity holds in all lattices of quasivarieties if and only if it is a consequence of join semidistributivity (Gorbunov [77], Theorem 5.2.9, based on [78]). See Section 4.2 for a discussion of these differences.

Our investigation of the structure of subquasivariety lattices begins with an analysis of the completely join irreducible and completely meet irreducible quasivarieties in $L_q(\mathcal{K})$, where $\mathcal{K}$ is locally finite and has finite type. Further, we will be concerned with the join and meet dependencies in $L_q(\mathcal{K})$. These results can be turned into algorithms for working with concrete locally finite quasivarieties of finite type. We will illustrate this by considering in depth a moderately complicated quasivariety of unary algebras, plus quasivarieties of 1-unary algebras and pure unary relational structures. When the calculations become involved or tedious, we turn to the Universal Algebra Calculator to aid in the analysis.

The main results of this monograph can be summarized as follows. We begin with an organized presentation of the elementary, well-known results (up through Theorem 2.13), and proceed from there. Let $\mathcal{K}$ be a locally finite quasivariety of finite type.

- The quasivariety $\langle \mathbf{T} \rangle$ generated by a finite algebra in $\mathcal{K}$ is compact in $L_q(\mathcal{K})$ (Lemma 2.4).
- The lattice $L_q(\mathcal{K})$ is both algebraic and dually algebraic (Theorem 2.5).
- A quasivariety is completely join irreducible in $L_q(\mathcal{K})$ if and only if it is generated by a single finite quasicritical algebra (Theorem 2.8).
- Every compact quasivariety has a canonical join representation in $L_q(\mathcal{K})$ (Theorem 2.13).

- For any finite algebra $\mathbf{T} \in \mathcal{K}$, we explicitly construct a finite set $\mathcal{E}(\mathbf{T})$ of quasi-equations such that, for any subquasivariety $\mathcal{Q} \leq \mathcal{K}$, we have $\mathbf{T} \notin \mathcal{Q}$ if and only if $\mathcal{Q}$ satisfies ε for some $\varepsilon \in \mathcal{E}(\mathbf{T})$ (Theorem 2.26).
- Every completely meet irreducible quasivariety in $L_q(\mathcal{K})$ is defined by a quasi-equation $\varepsilon \in \mathcal{E}(\mathbf{T})$ for some finite quasicritical algebra $\mathbf{T}$ (Theorem 2.30). However, not every such quasivariety need be meet irreducible.
- For any finite algebra $\mathbf{T}$ and $\varepsilon \in \mathcal{E}(\mathbf{T})$, there is a finite set of finite algebras $\mathbf{U}_1, \dots, \mathbf{U}_k$ such that an algebra $\mathbf{A} \in \mathcal{K}$ satisfies ε if and only if $\mathbf{A}$ omits each $\mathbf{U}_j$ as a subalgebra (Theorem 3.4).
- A quasivariety $\mathcal{Q} \leq \mathcal{K}$ is finitely based relative to $\mathcal{K}$ if and only if there exist finitely many quasicritical algebras $\mathbf{V}_1, \dots, \mathbf{V}_m$ such that for any algebra $\mathbf{A} \in \mathcal{K}$, we have $\mathbf{A} \in \mathcal{Q}$ if and only if $\mathbf{A}$ omits each $\mathbf{V}_j$ as a subalgebra (Theorem 3.10).
- $L_q(\mathcal{K})$ is a fermentable lattice (Corollary 4.7).

A principal tool in the analysis is reflection congruences, which are discussed in Section 2.2. A second tool is characterizing satisfaction of quasi-equations by omitted subalgebras, which is done in Section 2.3.

From our point of view, though, the algorithms that enable us to prove these results are as exciting as the consequences themselves. In Chapter 5 we apply the method to a variety $\mathcal{M}$ of 2-unary algebras with 0. The small quasicritical algebras, the quasivarieties they generate, and the relations between them in $L_q(\mathcal{M})$ are determined. The variety $\mathcal{M}$ in itself has no particular importance, but it is complicated enough to be interesting and provide a good test for the algorithms.

Then in Chapter 6 we apply the method to 1-unary algebras satisfying $f^r x \approx f^s x$. This yields a straightforward derivation of results of Kartashov [96, 97, 98, 99, 102]. Finally, in Chapter 7 we consider quasivarieties of pure unary relational structures.

The authors would like to thank colleagues who collaborated on various parts of this project. Ralph Freese added the test for quasicriticality to the Universal Algebra Calculator. David Casperson, Jesse Mason, and Brian Schaan worked with us on unary algebras with a 2-element range. We worked with Joel Adler on quasivarieties of 1-unary algebras, and with Kira Adaricheva on unary relational structures. The section on pseudoquasivarieties incorporates joint work with Steffen Lempp and Kira Adaricheva. Joy Nishida has been involved in studying the closely related topic of congruence lattices of semilattices with operators. Brian Davey, Lucy Ham, Marcel Jackson, and Tomasz Kowalski from the algebra seminar at Latrobe University provided a careful critique of an earlier version of the book. All these folk have contributed to the mathematics, and likewise to our enjoyment, of this investigation.

1.2. Background

Let us informally review the basic universal algebra that will be used in the sequel.

The *type* of a class of algebras or structures consists of their operation symbols and relation symbols, with their respective arities. The class has *finite type* if it has only finitely many such symbols, each of finite arity. In general, a *structure* may have both operations and relations. Structures with only operations are termed *algebras*, while those with only relations are called *pure relational structures*.

Note to the Reader: Up until Chapter 7, our results are stated in terms of algebras, rather than more general structures. This is perhaps more familiar territory for most readers. But the results are true for structures as well, and the changes required for this adaptation are straightforward.

Likewise, in this monograph we work in languages that contain an equality relation $\approx$. There are times when it is useful to consider implicational classes in languages without equality; see, for example, [17, 51, 132, 139].

Our notation from universal algebra is standard. In particular, given a structure $\mathbf{T}$, then $\mathrm{Sg}(X)$ denotes the substructure generated by a subset $X \subseteq T$, and $\mathrm{Cg}(Y)$ denotes the congruence generated by a set of pairs $Y \subseteq T^2$. We write $\mathbf{S} \leq \mathbf{T}$ to mean that $\mathbf{S}$ is isomorphic to a substructure of $\mathbf{T}$, that is, there is an embedding $h : \mathbf{S} \to \mathbf{T}$. Subdirect products are denoted by $\mathbf{T} \leq_s \mathbf{U}_1 \times \cdots \times \mathbf{U}_k$, meaning that $\mathbf{T} \leq \mathbf{U}_1 \times \cdots \times \mathbf{U}_k$ and the projection maps $\pi_i h$ are surjective.

Terms in a set of variables X are the smallest collection such that

(1) each $x \in X$ is a term;
(2) if $t_1, \ldots, t_m$ are terms and f is an operation symbol of arity m in the type, then $f(t_1, \ldots, t_m)$ is a term.

An *atomic formula* is an expression of the form $s \approx t$ where s, t are terms, or $R(s_1, \ldots, s_n)$ where $s_1, \ldots, s_n$ are terms and R is a relation symbol in the type.

A *quasi-equation* (or *definite Horn sentence*) is an implication of the form $\alpha_1 \& \ldots \& \alpha_k \to \beta$ where $\alpha_1, \ldots, \alpha_k$, β are atomic formulas. The antecedent is allowed to be empty ($k = 0$), so that atomic formulas themselves are quasi-equations.

Examples of quasi-equations from the language of lattices include distributivity and join semidistributivity:

$$x \wedge (y \vee z) \approx (x \wedge y) \vee (x \wedge z)$$
$$x \vee y \approx x \vee z \to x \vee y \approx x \vee (y \wedge z).$$

An example from the language of groups would be $x^2 \approx 1 \to xy \approx yx$, saying that transpositions are central.

We say that a structure **S** *satisfies* a quasi-equation φ if the implication is true for every substitution $\sigma : X \to S$ of elements of S for the variables. We say that **S** satisfies a collection Φ of quasi-equations if **S** satisfies every $\varphi \in \Phi$.

The quasi-equations satisfied by a class $\mathcal{K}$ of structures are often referred to as its *quasi-identities*.

A class of structures $\mathcal{Q}$ is a *quasivariety* if there is a set of quasi-equations Φ such that $\mathcal{Q}$ is the set of all structures (of the given type) that satisfy Φ. The classic characterization of quasivarieties uses the class operators: for a collection $\mathcal{X}$ of structures of the same type,

- $\mathbb{H}(\mathcal{X})$ denotes all homomorphic images of structures in $\mathcal{X}$,
- $\mathbb{S}(\mathcal{X})$ denotes all isomorphic copies of substructures of structures in $\mathcal{X}$,
- $\mathbb{P}(\mathcal{X})$ denotes all direct products of structures in $\mathcal{X}$,
- $\mathbb{U}(\mathcal{X})$ denotes all ultraproducts of structures in $\mathcal{X}$,
- $\mathbb{R}(\mathcal{X})$ denotes all reduced products of structures in $\mathcal{X}$.

Note that $\mathbb{P}(\mathcal{X})$ and $\mathbb{R}(\mathcal{X})$ include the empty product, which is a 1-element structure with all possible relations of the type holding. The classes $\mathbb{H}(\mathcal{X})$, $\mathbb{S}(\mathcal{X})$, etc. are taken to be closed under isomorphism, by convention.

THEOREM 1.1. *The following are equivalent for a class* $\mathcal{Q}$ *of structures of the same type.*

(1) $\mathcal{Q}$ *is a quasivariety.*
(2) $\mathcal{Q}$ *is closed under* $\mathbb{S}$, $\mathbb{P}$, *and* $\mathbb{U}$.
(3) $\mathcal{Q} = \mathbb{SPU}(\mathcal{Q})$.
(4) $\mathcal{Q} = \mathbb{SR}(\mathcal{Q})$.
(5) $\mathcal{Q} = \mathbb{SUP}(\mathcal{Q})$.

Let us denote the quasivariety generated by $\mathcal{X}$ as $\mathbb{Q}(\mathcal{X})$.

COROLLARY 1.2. *If* $\mathcal{X}$ *is a collection of structures of the same type, then*

$$\mathbb{Q}(\mathcal{X}) = \mathbb{SR}(\mathcal{X}) = \mathbb{SPU}(\mathcal{X}) = \mathbb{SUP}(\mathcal{X}) .$$

Theorem 1.1 is Corollary 2.3.4 in Gorbunov's book [77], based on Mal'cev [124], and incorporating ideas of Łoś (see [44]), Mal'cev [121], Horn [84], Chang and Morel [45] and Frayne, Morel, and Scott [66]. Our Lemma 2.2 gives part of the proof.

Of particular importance to us is the case when $\mathcal{X}$ consists of finitely many structures, say $\mathcal{X} = \{\mathbf{X}_1, \dots, \mathbf{X}_k\}$. In that case, any ultraproduct of members of $\mathcal{X}$ is isomorphic to some $\mathbf{X}_j$. This has the following consequence.

COROLLARY 1.3. *If* $\mathcal{X}$ *is a finite set of structures, then* $\mathbb{Q}(\mathcal{X}) = \mathbb{SP}(\mathcal{X})$.

If Φ is a set of atomic formulas, then the class of all models of Φ is called a *variety*. In that case we have the familiar characterization of Birkhoff [31]: A class $\mathcal{V}$ of structures is a variety if and only if $\mathbb{HSP}(\mathcal{V}) = \mathcal{V}$. The variety generated by a collection $\mathcal{X}$ of structures is denoted $\mathbb{V}(\mathcal{X})$.

Let $\mathcal{Q}$ be a quasivariety and $\mathbf{T} \in \mathcal{Q}$. Since quasivarieties need not be closed under homomorphic images, for a congruence $\theta \in \mathrm{Con}\,\mathbf{T}$ the quotient $\mathbf{T}/\theta$ may or may not be in $\mathcal{Q}$. Those congruences θ such that $\mathbf{T}/\theta \in \mathcal{Q}$ are called $\mathcal{Q}$-*congruences*. The $\mathcal{Q}$-congruences of $\mathbf{T}$ form a complete meet subsemilattice of $\mathrm{Con}\,\mathbf{T}$, denoted $\mathrm{Con}_{\mathcal{Q}}\,\mathbf{T}$.

If $\mathcal{Q}$ is a quasivariety and X any set, then there is a *free structure* $\mathcal{F}_{\mathcal{Q}}(X)$ with the properties

- X generates $\mathcal{F}_{\mathcal{Q}}(X)$;
- $\mathcal{F}_{\mathcal{Q}}(X) \in \mathcal{Q}$;
- if $\mathbf{T} \in \mathcal{Q}$, then any map $h_0 : X \to T$ extends to a homomorphism $h : \mathcal{F}_{\mathcal{Q}}(X) \to \mathbf{T}$.

For any collection $\mathcal{C}$ of structures, the free structure is constructed as a subdirect product of substructures of members of $\mathcal{C}$. Thus if $\mathcal{V}$ is the variety $\mathbb{HSP}(\mathcal{Q})$, then $\mathcal{F}_{\mathcal{V}}(X) = \mathcal{F}_{\mathcal{Q}}(X)$. The difference is that every homomorphic image of the free structure lies in $\mathcal{V}$, but not necessarily in $\mathcal{Q}$.

A quasivariety $\mathcal{Q}$ is *locally finite* if every finitely generated structure in $\mathcal{Q}$ is finite. This is equivalent to the free structure $\mathcal{F}_{\mathcal{Q}}(n)$ being finite for every positive integer n.

A quasivariety is *finitely generated* if it is generated by a finite collection of finite structures. It is not hard to see that every finitely generated quasivariety is locally finite. Proofs can be found in the universal algebra textbooks of Bergman [27] (Theorem 3.49), Burris and Sankappanavar [39] (Theorem II.10.16), or McKenzie, McNulty, and Taylor [119] (Theorem 4.99). These theorems are stated for varieties, but since $\mathcal{F}_{\mathcal{V}(\mathcal{X})}(Y) = \mathcal{F}_{\mathbb{Q}(\mathcal{X})}(Y)$ for any set of structures $\mathcal{X}$ and set Y, a quasivariety is locally finite if and only if the variety it generates is locally finite.

A locally finite quasivariety need not be finitely generated. The example of pseudocomplemented distributive lattices (or p-algebras) is given in Bergman [27], Theorem 4.55. Monadic algebras provide a similar example, given as Exercises 4.10.3 and 4.11.15 in McKenzie, McNulty, and Taylor [119].

For a quasivariety $\mathcal{K}$, let $\mathrm{L}_q(\mathcal{K})$ denote the lattice of quasivarieties contained in $\mathcal{K}$, ordered by containment. That is, $\mathcal{P} \leq \mathcal{Q}$ in $\mathrm{L}_q(\mathcal{K})$ if every structure in $\mathcal{P}$ is also in $\mathcal{Q}$, and every structure in $\mathcal{Q}$ is in $\mathcal{K}$. Likewise, for a variety $\mathcal{V}$, let $\mathrm{L}_v(\mathcal{V})$ denote the lattice of varieties contained in $\mathcal{V}$. If $\mathcal{V}$ is the variety and $\mathcal{K}$ the quasivariety generated by a finite structure, then the lattices $\mathrm{L}_v(\mathcal{V})$ and $\mathrm{L}_q(\mathcal{K})$ may be finite or infinite, in either order.

Consider the variety $\mathbb{V}(\mathbf{A})$ and the quasivariety $\mathbb{Q}(\mathbf{A})$ generated by a finite algebra $\mathbf{A}$. Then of course $\mathbb{Q}(\mathbf{A}) \subseteq \mathbb{V}(\mathbf{A})$, but it can easily happen that $\mathrm{L}_v(\mathbb{V}(\mathbf{A}))$ is finite, while $\mathrm{L}_q(\mathbb{Q}(\mathbf{A}))$ is infinite. In particular, if $\mathbf{A}$ lies in a congruence distributive variety, then $\mathbb{V}(\mathbf{A})$ will have only finitely many subvarieties, by Jónsson's Lemma [91]. But for the same algebra, the quasivariety $\mathbb{Q}(\mathbf{A})$ can have uncountably many subquasivarieties. A classical example of this phenomenon is the result of Grätzer and Lakser that the quasivariety generated by the 8-element lattice $\mathbf{M}_{3,3}$ contains continuum many subquasi-varieties [82].

On the other hand, when $\mathbb{Q}(\mathbf{A}) \subset \mathbb{V}(\mathbf{A})$, it is possible to have $\mathbb{L}_q(\mathbb{Q}(\mathbf{A}))$ finite and $\mathbb{L}_v(\mathbb{V}(\mathbf{A}))$ infinite. Let $\mathbf{N}_2^1$ denote the 3-element semigroup on the set $\{a, 0, 1\}$ with $a^2 = 0$, the element 0 a zero and 1 a unit. Then Carlisle [41, 62] showed that $|\mathbb{L}_v(\mathbb{V}(\mathbf{N}_2^1))| = \aleph_0$, while straightforward calculations using Corollary 2.29 yield that $|\mathbb{L}_q(\mathbb{Q}(\mathbf{N}_2^1))| = 5$.

In a probabilistic sense, almost all finite algebras are quasiprimal, by a theorem of Murskiĭ [129]. For a finite quasiprimal algebra $\mathbf{B}$, it happens that $\mathbb{Q}(\mathbf{B}) = \mathbb{V}(\mathbf{B})$, and $\mathbb{L}_q(\mathbb{Q}(\mathbf{B})) = \mathbb{L}_v(\mathbb{V}(\mathbf{B}))$ is a finite distributive lattice. The classical theory of these well-behaved algebras and the quasivarieties they generate is surveyed in Section A.6 of the Appendix.

A variety is *finitely based* if it is determined by finitely many atomic formulas. Let $\mathcal{V}$ be a variety and $\mathcal{K}$ a quasivariety of the same type. If $\mathcal{V} \leq \mathcal{K}$, then $\mathcal{V}$ is *finitely based relative to* $\mathcal{K}$ if $\mathcal{V}$ is determined by finitely many atomic formulas in addition to the axioms of $\mathcal{K}$. Likewise, a quasivariety is *finitely based* if it is determined by finitely many quasi-equations, and similarly for a subquasivariety $\mathcal{Q} \leq \mathcal{K}$ being *finitely based relative to* $\mathcal{K}$. Of course, when $\mathcal{K}$ is itself finitely based (as often is the case), these notions are the same.

Let us quickly review the better-known finite basis results for varieties, following largely the survey in Maróti and McKenzie [120]. Highlights include the Oates-Powell theorem that the variety generated by a finite group is finitely based [140], Kruse's theorem that the variety generated by a finite ring is finitely based [111], and Baker's theorem that a finitely generated congruence distributive variety is finitely based [24]. McKenzie proved that a finitely generated congruence modular variety with a cardinal bound on the size of its subdirectly irreducible members is finitely based [117], and Willard proved the corresponding result for congruence meet semidistributive varieties [169].

There are many examples of finitely generated varieties that are not finitely based. A couple of the originals are Lyndon's 7-element algebra [114] and Murskiĭ's 3-element groupoid [128]. Bryant showed that a finite group with a distinguished element as a constant need not be finitely based [37].

Turning to quasivarieties, Ol'šanskiĭ showed that the quasi-identities of a finite group are finitely based if and only if every Sylow subgroup of the group is abelian [141]. Belkin in [26] proved that a finite ring has a finite basis for its quasi-identities if and only if the ring satisfies $x^3 \approx 0 \to x^2 \approx 0$ and $x^2 \approx y^2 \approx xyx \approx 0 \to xy \approx 0$. Sapir proved that the quasi-identities of a finite completely simple semigroup are finitely based if and only if it is a rectangular group (the direct product of a group and a rectangular band), all Sylow subgroups of which are abelian [148, 149].

Other examples of finitely generated quasivarieties that are not finitely based include a 10-element lattice (Belkin [26]) and an 18-element semigroup with an additional unary operation (Lawrence and Willard [112]).

Our examples in Chapter 5 of finitely based and non-finitely based quasivarieties are unary algebras with a constant 0 and a two-element range $\{0, 1\}$. A criterion for when such an algebra generates a finitely based quasivariety (Theorem 5.2 below) was given in Casperson *et al.* [43]. See also [30, 42, 85, 100]. In particular, these papers provide lots of examples of finitely based and non-finitely based quasivarieties.

A quasivariety $\mathcal{K}$ is *relatively congruence distributive* if the lattice of $\mathcal{K}$-congruences $\mathrm{Con}_{\mathcal{K}} \mathbf{T}$ is distributive for every $\mathbf{T} \in \mathcal{K}$. As an analogue to Baker's theorem for varieties, Pigozzi proved that every finitely generated relatively congruence distributive quasivariety has a finite basis of quasi-equations [143]. See Blok and Pigozzi [36], Dziobiak [56], Nurakunov and Stronkowski [138] for related results on relatively congruence distributive quasivarieties.

The relatively congruence distributive theorem was extended to finitely generated, relatively congruence meet semidistributive quasivarieties by Dziobiak, Maróti, McKenzie, and Nurakunov [60]. An earlier version is that $\mathbb{SP}(\mathcal{K})$ is finitely axiomatizable when $\mathcal{K}$ is a finite collection of finite algebras such that $\mathbb{SP}(\mathcal{K})$ has pseudocomplemented congruence lattices and $\mathbb{HS}(\mathcal{K}) \subseteq \mathbb{SP}(\mathcal{K})$.

Pigozzi conjectured that every finitely generated relatively congruence modular quasivariety is finitely based. The best progress to date is the result of Dziobiak, Maróti, McKenzie, and Nurakunov that if a finitely generated relatively congruence modular quasivariety also generates a congruence modular variety, then it is finitely based [60].

1.3. Review of Complete Lattices

This section contains a short review of some lattice theory that we will use. More complete accounts include Davey and Priestley [47], Grätzer [81], and Nation [131]. There is also a good summary in McKenzie, McNulty, and Taylor [119].

A *lattice* is a partially ordered set in which every pair of elements x, y has a greatest lower bound, denoted $x \wedge y$ and called the *meet*, and a least upper bound, denoted $x \vee y$ and called the *join*. A *complete lattice* is a partially ordered set in which every subset X has a greatest lower bound, denoted $\bigwedge X$, and a least upper bound, denoted $\bigvee X$. The least element of a complete lattice $\mathbf{L}$ is denoted by $0_{\mathbf{L}}$, and its greatest element by $1_{\mathbf{L}}$. By convention, $\bigvee \varnothing = 0_{\mathbf{L}}$ and $\bigwedge \varnothing = 1_{\mathbf{L}}$.

The *dual* of a partially ordered set $\mathbf{P} = \langle P, \leq \rangle$ is the ordered set $\mathbf{P}^d = \langle P, \leq^d \rangle$ where $a \leq^d b$ iff $a \geq b$. If $\mathbf{P}$ is a lattice or complete lattice, then so is $\mathbf{P}^d$, with the meet and join operations interchanged. For any lattice notion, there is a dual notion obtained by interchanging $\leq$ and $\geq$, $\wedge$ and $\vee$, $\bigwedge$ and $\bigvee$.

A *complete meet semilattice* is a partially ordered set $\mathbf{S}$ in which every subset X has a greatest lower bound $\bigwedge X$. This includes that $\mathbf{S}$ should have a greatest element, $1_{\mathbf{S}} = \bigwedge \varnothing$. The crucial observation is that a complete meet semilattice is in fact already a complete lattice, with its join given by

$$\bigvee X = \bigwedge \{s \in S : s \geq x \text{ for all } x \in X\}.$$

That is, the least upper bound of a set is the meet of all its upper bounds, as long as that meet exists.

This leads us directly to the notion of a closure system. A *closure system* is a complete meet subsemilattice of a complete lattice. The most important cases are closure systems $\mathbf{S}$ that consist of a collection of subsets of a set U with the property that $\mathbf{S}$ is closed under arbitrary intersections, and hence a complete meet subsemilattice of the power set lattice $\mathcal{P}(U)$. Closure systems always form a complete lattice.

Examples of closure systems include the following.

(1) The collection of all closed sets in a topological space.
(2) The collection of all substructures of a structure, denoted Sub $\mathbf{S}$.
(3) The collection of all congruence relations on a structure, denoted Con $\mathbf{S}$.
(4) For a fixed quasivariety $\mathcal{Q}$, the collection of all congruence relations θ on a structure $\mathbf{S}$ such that $\mathbf{S}/\theta \in \mathcal{Q}$, denoted $\mathrm{Con}_\mathcal{Q} \mathbf{S}$.

An element a of a lattice $\mathbf{L}$ is *meet irreducible* if $a = \bigwedge X$ for a finite nonempty set X implies $a \in X$. The element a is *completely meet irreducible* if $a = \bigwedge X$ for an arbitrary set $X \subseteq L$ implies $a \in X$. A completely meet irreducible element a in a complete lattice has a unique upper cover,

$$a^* = \bigwedge \{x \in L : x > a\}.$$

The notions of join irreducible, completely join irreducible, and the unique lower cover a_* are defined dually, interchanging meet and join.

An element p of a lattice $\mathbf{L}$ is *join prime* if $p \leq \bigvee X$ for a finite nonempty set X implies $p \leq x$ for some $x \in X$. The element p is *completely join prime* if $p \leq \bigvee X$ for an arbitrary set $X \subseteq L$ implies $p \leq x$ for some $x \in X$. Of course, *meet prime* and *completely meet prime* are defined dually. In a complete lattice, completely join prime and completely meet prime elements come in pairs: if p is completely join prime, then $q = \bigvee \{x \in L : x \not\geq p\}$ is completely meet prime, and L is the disjoint union of the filter $\uparrow p = \{x \in L : x \geq p\}$ and the ideal $\downarrow q = \{x \in L : x \leq q\}$.

An element c of a complete lattice $\mathbf{L}$ is *compact* if whenever $c \leq \bigvee X$ for a subset $X \subseteq L$, there is a finite subset $F \subseteq X$ such that $c \leq \bigvee F$. Clearly $0_\mathbf{L}$ is compact, and the join of finitely many compact elements is compact.

A complete lattice is *algebraic* if every element is a (possibly infinite) join of compact elements. That is, $\mathbf{L}$ is algebraic if for every $x \in L$ we have $x = \bigvee \{c \in L : c \leq x \text{ and } c \text{ is compact}\}$.

Most of the important lattices that arise in algebra are either algebraic or dually algebraic. Substructure lattices Sub $\mathbf{S}$, congruence lattices Con $\mathbf{S}$, and lattices of relative congruences $\mathrm{Con}_\mathcal{Q} \mathbf{S}$ are all algebraic. On the other hand, lattices of subvarieties $\mathrm{L}_v(\mathcal{V})$ and lattices of subquasivarieties $\mathrm{L}_q(\mathcal{Q})$

are dually algebraic (because these lattices are each dually isomorphic to the corresponding lattice of theories, which is algebraic).

Algebraic lattices have some strong structural properties. A lattice $\mathbf{L}$ is *weakly atomic* if whenever $a < b$ in $\mathbf{L}$, there exist elements u, v such that $a \leq u \prec v \leq b$. A complete lattice is *upper continuous* if for every chain D in $\mathbf{L}$ and every $a \in L$, $a \wedge \bigvee D = \bigvee_{d \in D}(a \wedge d)$. Equivalently, $a \wedge \bigvee D = \bigvee_{d \in D}(a \wedge d)$ for every nonempty up-directed set contained in $\mathbf{L}$.

THEOREM 1.4. *Let $\mathbf{L}$ be an algebraic lattice.*

(1) $\mathbf{L}$ *is weakly atomic.*

(2) $\mathbf{L}$ *is upper continuous.*

(3) *Every element of $\mathbf{L}$ is a meet of completely meet irreducible elements.*

Part (3) is particularly important for quasivarieties because, when applied to congruence lattices Con $\mathbf{S}$ or to relative congruence lattices $\mathrm{Con}_\mathcal{Q}\, \mathbf{S}$, it yields subdirect decompositions into subdirectly irreducible structures (Birkhoff [31]).

COROLLARY 1.5. (1) *Every structure is a subdirect product of subdirectly irreducible structures.*

(2) *Every structure in a quasivariety $\mathcal{Q}$ is a subdirect product of structures that are subdirectly irreducible within $\mathcal{Q}$.*

Structures that are in a quasivariety $\mathcal{Q}$ and have no proper subdirect decomposition with factors in $\mathcal{Q}$ are called $\mathcal{Q}$-*subdirectly irreducible*.

The *join semidistributive* law is the quasi-equation

$$(\mathrm{SD}_\vee) \qquad x \vee y \approx x \vee z \rightarrow x \vee y \approx x \vee (y \wedge z).$$

Lattices of subquasivarieties $\mathrm{L_q}(\mathcal{Q})$ are known to be dually algebraic and join semidistributive [77]. Indeed, since being dually algebraic implies lower continuity, $\mathrm{L_q}(\mathcal{Q})$ satisfies the more general join semidistributive law,

$$u \approx x \vee z_i \text{ for all } i \in I \text{ implies } u \approx x \vee \bigwedge_{i \in I} z_i.$$

Section A.3 of the Appendix sketches a proof that a subquasivariety lattice $\mathrm{L_q}(\mathcal{Q})$ has the even stronger Jónsson-Kiefer Property; see Corollary A.1.

CHAPTER 2

Structure of Lattices of Subquasivarieties

2.1. Completely Join Irreducible Quasivarieties

Remember that the lattice $L_q(\mathcal{K})$ of all subquasivarieties of a quasivariety $\mathcal{K}$ is dually algebraic and join semidistributive. The goal of this section is to characterize the completely join irreducible quasivarieties in $L_q(\mathcal{K})$. Most of the results in this section can be found in Section 5.1 of Gorbunov's book [77].

Let us begin with some general considerations. It is convenient to use $\langle \mathbf{T} \rangle$ to denote the quasivariety generated by an algebra $\mathbf{T}$ in $\mathcal{K}$, that is $\langle \mathbf{T} \rangle = \mathbb{Q}(\mathbf{T})$ for a single algebra $\mathbf{T}$. It will cause no confusion to also use $\langle \varepsilon \rangle$ to denote the quasivariety consisting of all algebras in $\mathcal{K}$ that satisfy a quasiequation ε.

Finite joins of quasivarieties have a simple description.

LEMMA 2.1. *Let $\mathcal{K}$ be any quasivariety. Consider subquasivarieties $\mathcal{Q}$, $\mathcal{R} \leq \mathcal{K}$ and an algebra $\mathbf{T} \in \mathcal{K}$. Then $\mathbf{T} \in \mathcal{Q} \vee \mathcal{R}$ if and only if $\mathbf{T}$ is a subdirect product of algebras in $\mathcal{Q}$ and $\mathcal{R}$.*

PROOF. If $\mathbf{T} \leq \mathbf{A} \times \mathbf{B}$ with $\mathbf{A} \in \mathcal{Q}$ and $\mathbf{B} \in \mathcal{R}$, then $\mathbf{T} \in \mathcal{Q} \vee \mathcal{R}$.

Conversely, assume $\mathbf{T} \in \mathcal{Q} \vee \mathcal{R} = \mathbb{SPU}(\mathcal{Q} \cup \mathcal{R})$. An ultraproduct of algebras in $\mathcal{Q} \cup \mathcal{R}$ is isomorphic to either an ultraproduct of algebras in $\mathcal{Q}$ or ultraproduct of algebras in $\mathcal{R}$, and both $\mathcal{Q}$ and $\mathcal{R}$ are closed under ultraproducts. Hence $\mathbf{T} \in \mathbb{SP}(\mathcal{Q} \cup \mathcal{R})$, wherefore it is a subdirect product of algebras in $\mathcal{Q}$ and $\mathcal{R}$. $\qquad\square$

Infinite joins of quasivarieties require more care. Recall Corollary 1.2 that for any class of structures $\mathcal{X}$, the quasivariety generated by $\mathcal{X}$ is given by
$$\mathbb{Q}(\mathcal{X}) = \mathbb{SR}(\mathcal{X}) = \mathbb{SPU}(\mathcal{X}) = \mathbb{SUP}(\mathcal{X}).$$
The characterization $\mathbb{Q}(\mathcal{X}) = \mathbb{SR}(\mathcal{X}) = \mathbb{SPU}(\mathcal{X})$ is a fundamental theorem of model theory. There is a complete proof in Chapter V of Burris and Sankappanavar [39]; see Theorem V.2.25. The equality $\mathbb{SR}(\mathcal{X}) = \mathbb{SPU}(\mathcal{X})$ just

J. Hyndman, J. B. Nation, *The Lattice of Subquasivarieties of a Locally Finite Quasivariety*, CMS Books in Mathematics, https://doi.org/10.1007/978-3-319-78235-5_2

reflects the fact that any filter on a boolean algebra (in this case the boolean algebra $\mathbf{2}^I$, where I is the index set for a direct product) is an intersection of ultrafilters.

One version of the characterization of quasivarieties in Theorem 1.1 and Corollary 1.2 may be unfamiliar to many readers, *viz.*, that $\mathbb{Q}(\mathcal{X}) = \mathbb{SUP}(\mathcal{X})$. Since that is the version we use in the proof of Lemma 2.4, and the proof in Gorbunov [77] is rather roundabout, let us give a direct, elementary proof that this is equivalent to one of the standard characterizations.

LEMMA 2.2. *If $\mathcal{X}$ is a collection of algebras of the same type, then* $\mathbb{SR}(\mathcal{X}) = \mathbb{SUP}(\mathcal{X})$.

PROOF. Since direct products and ultraproducts are special types of reduced products, and $\mathbb{R}^2(\mathcal{X}) = \mathbb{R}(\mathcal{X})$, it is clear that $\mathbb{SUP}(\mathcal{X}) \subseteq \mathbb{SR}(\mathcal{X})$.

For the reverse inclusion, consider an algebra $\mathbf{A} \in \mathbb{SR}(\mathcal{X})$. Then there exist a collection $\mathbf{B}_i$ ($i \in I$) of algebras in $\mathcal{X}$, a filter F on I, and an embedding $\gamma : \mathbf{A} \le (\prod_{i \in I} \mathbf{B}_i)/{\equiv_F}$. We may assume $\varnothing \notin F$, else $\mathbf{A}$ is trivial.

For each $a \in A$, choose a representative $\delta(a)$ of the $\equiv_F$-class $\gamma(a)$. Since γ is one-to-one, we have $\delta(a) \equiv_F \delta(c)$ if and only if $a = c$. If we use the notation $[\![x = y]\!]_S = \{i \in S : x_i = y_i\}$, this becomes $[\![\delta(a) = \delta(c)]\!]_I \in F$ iff $a = c$. The homomorphism property of γ can then be written as

$$[\![f(\delta(a_1), \ldots, \delta(a_k)) = \delta(f(a_1, \ldots, a_k))]\!]_I \in F$$

whenever f is a fundamental operation of arity k and $a_1, \ldots, a_k \in A$.

The next observation, however simple, is crucial.

LEMMA 2.3. *If $x, y \in \prod_{i \in I} B_i$ and $T \subseteq S = [\![x = y]\!]_I$, then $x|_T = y|_T$ in the direct product $\prod_{j \in T} \mathbf{B}_j$.*

Continuing with the proof of Lemma 2.2, let $\mathbf{D} = \prod_{S \in F}(\prod_{i \in S} \mathbf{B}_i)$. We describe an ultrafilter V on F and an embedding $\beta : \mathbf{A} \le \mathbf{D}/{\equiv_V}$. This will show that $\mathbf{A} \in \mathbb{SUP}(\mathcal{X})$.

Let V_0 be the collection of all sets $\downarrow S \cap F = \{T \in F : T \subseteq S\}$ with $S \in F$. Note that $(\downarrow S_1 \cap F) \cap (\downarrow S_2 \cap F) = \downarrow(S_1 \cap S_2) \cap F$, and $S_1 \cap S_2$ is in F when $S_1, S_2 \in F$. Thus V_0 is closed under finite intersections and does not contain the empty set. Hence V_0 can be extended to an ultrafilter V on F.

Let $\varepsilon : \mathbf{A} \to \mathbf{D}$ be defined by $\varepsilon(a)_S = \delta(a)|_S$, that is, $\varepsilon(a)$ is the vector whose S-th component is the projection of $\delta(a)$ onto $\prod_{i \in S} \mathbf{B}_i$. Then define $\beta(a) = \varepsilon(a)/{\equiv_V}$. We must show that β is an embedding, i.e., one-to-one and a homomorphism.

Let $a, c \in A$ and suppose $\beta(a) = \beta(c)$, so that $\varepsilon(a) \equiv_V \varepsilon(c)$. Thus $[\![\varepsilon(a) = \varepsilon(c)]\!]_F$ is in V; in particular, it is nonempty. Hence there exists an $S_0 \in F$ such that $\varepsilon(a)_{S_0} = \varepsilon(c)_{S_0}$, i.e., $\delta(a)|_{S_0} = \delta(c)|_{S_0}$. Therefore $[\![\delta(a) = \delta(c)]\!]_I \supseteq S_0$, whence it is in F and $a = c$.

It remains to show that

$$[\![f(\varepsilon(a_1), \ldots, \varepsilon(a_k)) = \varepsilon(f(a_1, \ldots, a_k))]\!]_F \in V$$

whenever f is a fundamental operation of arity k and $a_1, \ldots, a_k \in A$. But we know that

$$S_1 = [\![f(\delta(a_1), \ldots, \delta(a_k)) = \delta(f(a_1, \ldots, a_k))]\!]_I \in F.$$

Therefore, by Lemma 2.3, for every set T in $\downarrow S_1 \cap F$ we have that $f(\delta(a_1), \ldots, \delta(a_k))|_T = \delta(f(a_1, \ldots, a_k))|_T$. Thus $\downarrow S_1 \cap F$ is contained in $[\![f(\varepsilon(a_1), \ldots, \varepsilon(a_k)) = \varepsilon(f(a_1, \ldots, a_k))]\!]_F$, whence the latter is in V. $\qquad\square$

Recall from Corollary 1.3 that if $\mathcal{X}$ consists of finitely many finite algebras, then $\mathbb{Q}(\mathcal{X}) = \mathbb{SP}(\mathcal{X})$. In particular, if $\mathbf{T}$ is finite, then $\langle \mathbf{T} \rangle = \mathbb{SP}(\mathbf{T})$.

LEMMA 2.4. *Assume that $\mathcal{K}$ has finite type, and let $\mathbf{T}$ be a finite algebra in $\mathcal{K}$. Then $\langle \mathbf{T} \rangle$ is compact in the lattice of subquasivarieties of $\mathcal{K}$.*

PROOF. Let $\mathcal{H}_i$ for $i \in I$ be quasivarieties contained in $\mathcal{K}$, and suppose that $\mathbf{T}$ is in their join, which is the quasivariety generated by $\mathcal{X} = \bigcup_{i \in I} \mathcal{H}_i$. Then $\mathbf{T} \in \mathbb{SUP}(\mathcal{X})$, so that $\mathbf{T} \leq \mathbf{A}$ for some algebra $\mathbf{A}$ that is an ultraproduct of algebras $\mathbf{B}_j$ for $j \in J$, say $\mathbf{A} = \prod_{j \in J} \mathbf{B}_j / \equiv_U$, and each $\mathbf{B}_j$ is a product of algebras from $\mathcal{X}$. Since $\mathbf{T}$ is a finite algebra of finite type, there is a (long) first-order sentence saying that an algebra $\mathbf{B}$ contains a subalgebra isomorphic to $\mathbf{T}$. Since this holds in the ultraproduct, the set of indices j such that $\mathbf{T}$ can be embedded into the algebra $\mathbf{B}_j$ is in the ultrafilter U. In particular, it is nonempty, so $\mathbf{T}$ can be embedded into some $\mathbf{B}_{j_0}$. Now $\mathbf{B}_{j_0} = \prod_{k \in K} \mathbf{C}_k$ with each $\mathbf{C}_k \in \mathcal{X} = \bigcup_{i \in I} \mathcal{H}_i$. Again because $\mathbf{T}$ is finite, only finitely many k's are needed, so that $\mathbf{T} \leq \prod_{k \in F} \mathbf{C}_k$ for some finite subset $F \subseteq K$. But now only finitely many $\mathcal{H}_i$'s are involved, and thus $\langle \mathbf{T} \rangle$ is compact. $\qquad\square$

If follows of course that the quasivariety generated by finitely many finite algebras in $\mathcal{K}$ is compact in $\mathrm{L}_q(\mathcal{K})$. Mal'cev observed that, in general, a quasivariety $\mathcal{Q}$ is generated by a single algebra if and only if whenever $\mathbf{A}$, $\mathbf{B} \in \mathcal{Q}$ there exists $\mathbf{C} \in \mathcal{Q}$ such that both $\mathbf{A}$ and $\mathbf{B}$ are embeddable into $\mathbf{C}$; see [124, 59]. For example, if $\mathbf{A}$ and $\mathbf{B}$ each contains an element that is idempotent for all operations, then $\langle \mathbf{A} \rangle \vee \langle \mathbf{B} \rangle = \langle \mathbf{A} \times \mathbf{B} \rangle$. On the other hand, if $\mathbf{C}_2$ and $\mathbf{C}_3$ are 1-unary algebras that are cycles of length 2 and 3, respectively, then $\langle \mathbf{C}_2 \rangle \vee \langle \mathbf{C}_3 \rangle$ is not generated by a single algebra.

Easy examples show that a finitely generated quasivariety need not be compact in $\mathrm{L}_q(\mathcal{K})$ if $\mathcal{K}$ does not have finite type. In order not to disrupt the flow of thought, we defer these examples to Section 2.6 at the end of this chapter.

THEOREM 2.5. *If $\mathcal{K}$ is a locally finite quasivariety of finite type, then the lattice of quasivarieties $\mathrm{L}_q(\mathcal{K})$ is both algebraic and dually algebraic.*

PROOF. Every subquasivariety lattice $\mathrm{L}_q(\mathcal{K})$ is dually algebraic. Assuming that $\mathcal{K}$ is locally finite and has finite type, we use Lemma 2.4 to show that it is also algebraic.

Let $\mathcal{Q}_1$, $\mathcal{Q}_2$ be quasivarieties in $\mathcal{K}$ with $\mathcal{Q}_1 < \mathcal{Q}_2$. Then there is a quasi-equation μ that holds in $\mathcal{Q}_1$ but not in $\mathcal{Q}_2$. Hence there is a finitely generated algebra $\mathbf{T} \in \mathcal{Q}_2 \setminus \mathcal{Q}_1$. Since $\mathcal{K}$ is locally finite, $\mathbf{T}$ is finite. Then $\langle \mathbf{T} \rangle$ is a compact quasivariety below $\mathcal{Q}_2$ but not below $\mathcal{Q}_1$ in $L_q(\mathcal{K})$. $\square$

An argument only slightly more involved shows that finite algebras generate compact varieties in the lattice $L_v(\mathcal{V})$ of subvarieties of a locally finite variety $\mathcal{V}$ of algebras; see Theorem 2.44 in Section 2.6. Thus these lattices are also algebraic and dually algebraic.

Whenever $\mathbf{L}$ is a complete lattice and c is compact in $\mathbf{L}$, then by Zorn's Lemma there exist elements that are maximal with respect to being not above c. Any such element is necessarily completely meet irreducible. Applying this to $L_q(\mathcal{K})$ yields the following.

COROLLARY 2.6. *Let* $\mathbf{T}$ *be a finite algebra in a quasivariety* $\mathcal{K}$ *of finite type. For any quasivariety* $\mathcal{Q}$ *with* $\mathbf{T} \notin \mathcal{Q}$, *there exists a completely meet irreducible quasivariety maximal with respect to being above* $\mathcal{Q}$ *but not above* $\langle \mathbf{T} \rangle$ *in* $L_q(\mathcal{K})$.

For a finite algebra $\mathbf{T}$ in $\mathcal{K}$, let us define $\kappa(\mathbf{T})$ to be the set of all quasivarieties $\mathcal{N} \leq \mathcal{K}$ that are maximal with respect to the property that $\mathbf{T} \notin \mathcal{N}$. Corollary 2.6 guarantees that $\kappa(\mathbf{T})$ is nonempty, and we will see shortly that it is finite. Theorem 2.26 and the discussion following it show how we can determine which quasivarieties are in $\kappa(\mathbf{T})$.

A lattice is *spatial* when every element is a join of completely join irreducible elements. A lattice is *dually spatial* when every element is a meet of completely meet irreducible elements. Every algebraic lattice is dually spatial lattice. Our lattices $L_q(\mathcal{K})$ are both algebraic and dually algebraic, so we have this consequence.

COROLLARY 2.7. *If* $\mathcal{K}$ *is a locally finite quasivariety of finite type, then* $L_q(\mathcal{K})$ *is both spatial and dually spatial.*

Now we turn to identifying the completely join irreducible quasivarieties in $L_q(\mathcal{K})$.

A finite algebra $\mathbf{T}$ is *quasicritical* if $\mathbf{T}$ is not in the quasivariety generated by its proper subalgebras. The origins of the next theorem are lost in antiquity.

THEOREM 2.8. *Let* $\mathcal{K}$ *be a locally finite quasivariety of finite type. A quasivariety is completely join irreducible in* $L_q(\mathcal{K})$ *if and only if it is generated by a single finite quasicritical algebra.*

The proof employs a technical lemma that will be used again later. Recall that $\mathbb{Q}(\mathcal{X})$ denotes the quasivariety generated by a set $\mathcal{X}$ of algebras.

LEMMA 2.9. *Let* $\mathcal{K}$ *be a locally finite quasivariety of finite type. Let* $\mathbf{S}_1, \ldots, \mathbf{S}_m$ *be finite algebras in* $\mathcal{K}$, *and let* $\mathbf{T}$ *be a finite quasicritical algebra in* $\mathcal{K}$. *If* $\mathbb{Q}(\mathbf{S}_1, \ldots, \mathbf{S}_m) = \mathbb{Q}(\mathbf{T})$, *then* $\mathbf{T} \leq \mathbf{S}_j$ *for some* j.

PROOF. Recall that $\Delta = \Delta_{\mathbf{T}}$ denotes the least congruence on $\mathbf{T}$. The fact that $\mathbf{T} \in \mathbb{Q}(\mathbf{S}_1, \ldots, \mathbf{S}_m) = \mathbb{SP}(\mathbf{S}_1, \ldots, \mathbf{S}_m)$ means that $\mathbf{T}$ is a subdirect product of subalgebras of the algebras $\mathbf{S}_j$. Thus in $\mathrm{Con}_{\mathcal{K}} \mathbf{T}$ there is a decomposition $\Delta_{\mathbf{T}} = \bigwedge_{i \leq n} \varphi_i$ where for each i there exists j_i such that $\mathbf{T}/\varphi_i \leq \mathbf{S}_{j_i}$.

For each $i \leq n$, let $\mathbf{R}_i \cong \mathbf{T}/\varphi_i$. Since $\mathbf{R}_i \leq \mathbf{S}_{j_i}$ is in $\mathbb{Q}(\mathbf{T})$, we have $\mathbf{R}_i \leq \mathbf{T}^{\ell}$ for some ℓ. Thus in $\mathrm{Con}_{\mathcal{K}} \mathbf{T}$ there is a decomposition $\varphi_i = \bigwedge_k \psi_{ik}$ with $\mathbf{T}/\psi_{ik} \leq \mathbf{T}$ for each k.

So $\Delta_{\mathbf{T}} = \bigwedge_i \bigwedge_k \psi_{ik}$ with each $\mathbf{T}/\psi_{ik} \leq \mathbf{T}$. Since $\mathbf{T}$ is quasicritical, there is a pair (i_0, k_0) such that $\Delta_{\mathbf{T}} = \psi_{i_0 k_0}$. But then $\varphi_{i_0} = \bigwedge_k \psi_{i_0 k} = \Delta_{\mathbf{T}}$, whence $\mathbf{T} = \mathbf{T}/\Delta_{\mathbf{T}} \leq \mathbf{S}_{j_{i_0}}$, as desired. $\qquad\square$

The proof of the one direction of Theorem 2.8 is now readily finished. Let $\mathbf{T}$ be a finite quasicritical algebra in $\mathcal{K}$. The fact that $\langle \mathbf{T} \rangle$ is compact means that if $\langle \mathbf{T} \rangle$ were completely join reducible, then it would be join reducible. So assume that $\langle \mathbf{T} \rangle = \mathcal{Q}_1 \vee \cdots \vee \mathcal{Q}_m$ in $\mathrm{L_q}(\mathcal{K})$. Then $\mathbf{T}$ can be written as a subdirect product, $\mathbf{T} \leq_s \mathbf{S}_1 \times \cdots \times \mathbf{S}_m$ with each $\mathbf{S}_i \in \mathcal{Q}_i$. Since each $\mathcal{Q}_i \leq \langle \mathbf{T} \rangle$, it follows that $\mathbb{Q}(\mathbf{S}_1, \ldots, \mathbf{S}_m) = \langle \mathbf{T} \rangle$. By Lemma 2.9, $\mathbf{T} \leq \mathbf{S}_j$ for some j. Then $\langle \mathbf{T} \rangle \leq \langle \mathbf{S}_j \rangle \leq \mathcal{Q}_j \leq \langle \mathbf{T} \rangle$, whence $\mathcal{Q}_j = \langle \mathbf{T} \rangle$. Thus $\langle \mathbf{T} \rangle$ is completely join irreducible.

For the converse, we use this lemma of Mal'cev [126]:

LEMMA 2.10. *Every algebra can be embedded into an ultraproduct of its finitely generated subalgebras.*

It remains to show that if $\mathcal{Q}$ is a completely join irreducible quasivariety, then $\mathcal{Q}$ is generated by a quasicritical algebra. However, $\mathcal{Q}$ is generated by any algebra $\mathbf{T}$ that is in $\mathcal{Q}$ but not in its lower cover $\mathcal{Q}_*$. By Mal'cev's lemma, $\mathbf{T}$ can be chosen to be finitely generated, and hence finite, since $\mathcal{K}$ is locally finite. Such an algebra of minimal size must be quasicritical. This completes the proof of Theorem 2.8.

Because the lattice of quasivarieties is dually algebraic, every quasivariety is determined by the completely join irreducible quasivarieties it contains, or equivalently for $\mathcal{K}$ locally finite, by the quasicritical algebras it contains.

Remark. We have not defined *quasicritical* for infinite algebras. The definition should of course be one that enables us to prove the analogue of Theorem 2.8 for general quasivarieties.

Applying Lemma 2.9 with $m = 1$ and $\mathbf{S}$, $\mathbf{T}$ both quasicritical yields an important consequence.

THEOREM 2.11. *Let $\mathcal{K}$ be a locally finite quasivariety of finite type. If $\mathbf{S}$ and $\mathbf{T}$ are distinct finite quasicritical algebras in $\mathcal{K}$, then $\langle \mathbf{S} \rangle \neq \langle \mathbf{T} \rangle$.*

Let us pause to examine how the preceding results relate to the structure of $\mathrm{L_q}(\mathcal{K})$. First, it follows from Mal'cev's lemma that a compact quasivariety can be generated by finitely many, finitely generated algebras. Thus when $\mathcal{K}$ is locally finite of finite type, the compact quasivarieties in $\mathrm{L_q}(\mathcal{K})$ are generated by finitely many, finite algebras. In fact, let us show that there is a canonical such representation in this case.

Recall that for subsets A, B of a lattice L, we say that A *refines* B, written $A \ll B$, if for every $a \in A$ there exists $b \in B$ with $a \leq b$. We say that an element $c \in L$ has a *canonical join representation* if there is a finite set $D \subseteq L$ such that $c = \bigvee D$ irredundantly, and whenever $c = \bigvee E$ with E finite, then $D \ll E$. The canonical join representation D necessarily consists of an antichain of join irreducible elements. The *canonical* part is that no element of D can be omitted or replaced by a set of lower elements in the lattice.

Dual refinement, written $A \gg B$, and *canonical meet representation* are defined dually. Note that, because of the quantification, $A \ll B$ is not the same as $B \gg A$.

Standard lattice theory arguments give the canonical join representation for compact elements (Gorbunov [75]).

THEOREM 2.12. *Let c be a compact element in a complete, lower continuous, join semidistributive lattice* **L**. *Then c has a canonical join representation in* **L**.

Of course, the canonical joinands of a compact element must be completely join irreducible.

Theorem 2.12 applies to $L_q(\mathcal{Q})$ for any quasivariety $\mathcal{Q}$, because it is join semidistributive and dually algebraic, with the latter property implying lower continuity.

THEOREM 2.13. *For any quasivariety $\mathcal{Q}$, the compact elements of $L_q(\mathcal{Q})$ have a canonical join representation.*

When $\mathcal{K}$ is a locally finite quasivariety of finite type, we know exactly what the compact quasivarieties in $L_q(\mathcal{K})$ are, *viz.*, those generated by finitely many finite quasicritical algebras. In Section 4.1, we will refine this observation considerably by showing that $L_q(\mathcal{K})$ is fermentable; see Corollary 4.7.

It is hard to see exactly what this decomposition says about the structure of a finite algebra $\mathbf{T} \in \mathcal{K}$. So let us look at a direct proof of the canonical form for the quasivariety generated by a single finite algebra. The technical lemma that does all the work requires some setup.

Let $\mathbf{T}$ be a finite algebra in a locally finite quasivariety $\mathcal{K}$ of finite type. Let

$$\mathcal{S} = \{\theta \in \mathrm{Con}_{\mathcal{K}} \, \mathbf{T} : \mathbf{T}/\theta \leq \mathbf{T}\},$$

i.e., $\mathcal{S}$ consists of those $\mathcal{K}$-congruences on $\mathbf{T}$ such that the factor algebra is isomorphic to a subalgebra of $\mathbf{T}$. Of course, $\Delta_{\mathbf{T}} \in \mathcal{S}$. We say that $\Phi \subseteq \mathrm{Con}_{\mathcal{K}} \, \mathbf{T}$ is a *maximal meet representation* of $\Delta_{\mathbf{T}}$ in $\mathcal{S}$ if

   (i) $\Phi \subseteq \mathcal{S}$,
   (ii) $\Delta_{\mathbf{T}} = \bigwedge \Phi$,
   (iii) if $\Psi \subseteq \mathcal{S}$, $\Delta_{\mathbf{T}} = \bigwedge \Psi$ and $\Psi \gg \Phi$, then $\Phi \subseteq \Psi$.

(This is the standard definition of a maximal meet representation, relativized to meets where the meetands are required to lie in S.)

LEMMA 2.14. *Let* $\mathbf{T}$ *be a finite algebra in a locally finite quasivariety* $\mathcal{K}$ *of finite type, and define* S *as above. Let* $\Phi = \{\varphi_1, \ldots, \varphi_m\}$ *be a maximal meet representation of* $\Delta_{\mathbf{T}}$ *in* S. *For* $i \leq m$, *let* $\mathbf{S}_i \cong \mathbf{T}/\varphi_i$. *Then*

(1) $\langle \mathbf{T} \rangle = \langle \mathbf{S}_1 \rangle \vee \cdots \vee \langle \mathbf{S}_m \rangle$,
(2) *whenever* $\langle \mathbf{T} \rangle = \mathcal{Q}_1 \vee \cdots \vee \mathcal{Q}_n$ *in* $L_q(\mathcal{K})$, *then*

$$\{\langle \mathbf{S}_1 \rangle, \ldots, \langle \mathbf{S}_m \rangle\} \ll \{\mathcal{Q}_1, \ldots, \mathcal{Q}_n\}.$$

PROOF. The first part holds because $\mathbf{T} \leq \mathbf{S}_1 \times \cdots \times \mathbf{S}_m$ and $\mathbf{S}_i \leq \mathbf{T}$ for all i. For part (2), we must show that if $\langle \mathbf{T} \rangle = \mathcal{Q}_1 \vee \cdots \vee \mathcal{Q}_n$, then for all $\mathbf{S}_i$ there exists $\mathcal{Q}_j$ such that $\langle \mathbf{S}_i \rangle \leq \mathcal{Q}_j$, i.e., $\mathbf{S}_i \in \mathcal{Q}_j$.

Now $\langle \mathbf{T} \rangle \leq \mathcal{Q}_1 \vee \cdots \vee \mathcal{Q}_n$ means that $\mathbf{T} \leq_s \mathbf{R}_1 \times \cdots \times \mathbf{R}_n$ with each $\mathbf{R}_j \in \mathcal{Q}_j$. That is, there exist congruences ψ_j on $\mathbf{T}$ such that $\Delta_{\mathbf{T}} = \bigwedge_j \psi_j$ and $\mathbf{T}/\psi_j \cong \mathbf{R}_j$.

Meanwhile, $\mathcal{Q}_j \leq \langle \mathbf{T} \rangle$ means that each $\mathbf{R}_j \leq \mathbf{T}^{n_j}$ for some n_j, implying that $\mathbf{R}_j \leq_s \mathbf{U}_{j1} \times \cdots \times \mathbf{U}_{jn_j}$ with each $\mathbf{U}_{jk} \leq \mathbf{T}$. (Note though, the algebras $\mathbf{U}_{jk}$ need not be in $\mathcal{Q}_j$.) Thus in each $\mathrm{Con}_{\mathcal{K}} \mathbf{R}_j$ there are congruences θ_{jk} such that $\Delta_{\mathbf{R}_j} = \bigwedge_k \theta_{jk}$ and $\mathbf{R}_j/\theta_{jk} \cong \mathbf{U}_{jk}$. By the Isomorphism Theorems, there are congruences $\hat{\theta}_{jk} \in \mathrm{Con}_{\mathcal{K}} \mathbf{T}$ such that $\psi_j = \bigwedge_k \hat{\theta}_{jk}$ and $\mathbf{T}/\hat{\theta}_{jk} \cong \mathbf{U}_{jk}$. Hence $\Delta_{\mathbf{T}} = \bigwedge_j \bigwedge_k \hat{\theta}_{jk}$.

For each $i \leq m$, consider the restrictions $\hat{\theta}_{jk}|_{\mathbf{S}_i}$. Clearly we have $\Delta_{\mathbf{S}_i} = \bigwedge_j \bigwedge_k \hat{\theta}_{jk}|_{\mathbf{S}_i}$. Again by the Isomorphism Theorems, there are congruences $\tilde{\theta}_{jk} \in \mathrm{Con}_{\mathcal{K}} \mathbf{T}$ such that $\varphi_i = \bigwedge_j \bigwedge_k \tilde{\theta}_{jk}$ and $\mathbf{T}/\tilde{\theta}_{jk} \cong \mathbf{S}_i/\hat{\theta}_{jk}|_{\mathbf{S}_i}$. Thus

$$\mathbf{T}/\tilde{\theta}_{jk} \cong \mathbf{S}_i/\hat{\theta}_{jk}|_{\mathbf{S}_i} \leq \mathbf{T}/\hat{\theta}_{jk} \cong \mathbf{U}_{jk} \leq \mathbf{T}$$

whence each $\tilde{\theta}_{jk}$ is in S.

Let $\tilde{\Theta}$ denote the collection on those congruences $\tilde{\theta}_{jk}$. Then $\tilde{\Theta} \subseteq S$, $\Delta_{\mathbf{T}} = \bigwedge \tilde{\Theta}$, and $\tilde{\Theta} \gg \Phi$. By the maximality of Φ, this implies $\Phi \subseteq \tilde{\Theta}$. That is, for each $i \leq m$ there exists a pair (j_i, k_i) such that $\varphi_i = \tilde{\theta}_{j_i, k_i}$. A fortiori, $\varphi_i = \bigwedge_k \tilde{\theta}_{j_i k}$.

Recall that $\psi_{j_i} = \bigwedge_k \hat{\theta}_{j_i k}$, whence $\psi_{j_i}|_{\mathbf{S}_i} = \bigwedge_k \hat{\theta}_{j_i k}|_{\mathbf{S}_i}$. Again using the Isomorphism Theorems, the congruence $\tilde{\psi}_{j_i} \geq \varphi_i$ such that $\mathbf{T}/\tilde{\psi}_{j_i} \cong \mathbf{S}_i/\psi_{j_i}|_{\mathbf{S}_i}$ satisfies $\tilde{\psi}_{j_i} = \bigwedge_k \tilde{\theta}_{j_i k}$. Thus $\tilde{\psi}_{j_i} = \varphi_i$.

Now

$$\mathbf{T}/\tilde{\psi}_{j_i} \cong \mathbf{S}_i/\psi_{j_i}|_{\mathbf{S}_i} \leq \mathbf{T}/\psi_{j_i} \cong \mathbf{R}_{j_i} \in \mathcal{Q}_{j_i},$$

whence $\mathbf{S}_i \cong \mathbf{T}/\varphi_i \in \mathcal{Q}_{j_i}$, as desired.   □

While we will see more complex instances of Lemma 2.14 in later chapters, the abelian group $\mathbf{Z}_p^2$, with congruence lattice $\mathbf{M}_{p+1}$, provides a simple example. The maximal meet representation Φ of Δ in S need not be unique, and the factors $\mathbf{S}_i = \mathbf{T}/\varphi_i$ need not be distinct. Indeed, two quasivarieties

$\langle \mathbf{S}_i \rangle$ and $\langle \mathbf{S}_j \rangle$ could be equal or comparable. But the maximal quasivarieties in the set $\{\langle \mathbf{S}_1 \rangle, \ldots, \langle \mathbf{S}_k \rangle\}$ *are* uniquely determined, and they are the members of the canonical join representation of $\langle \mathbf{T} \rangle$.

THEOREM 2.15. *Let* $\mathbf{T}$ *be a finite algebra in a locally finite quasivariety* $\mathcal{K}$ *of finite type, and let*

$$\mathcal{S} = \{\theta \in \mathrm{Con}_{\mathcal{K}} \mathbf{T} : \mathbf{T}/\theta \leq \mathbf{T}\},$$

For $i \leq k$, *let* $\mathbf{S}_i \cong \mathbf{T}/\varphi_i$. *Then the maximal quasivarieties in the set* $\{\langle \mathbf{S}_1 \rangle, \ldots, \langle \mathbf{S}_k \rangle\}$ *are the members of the canonical join representation of* $\langle \mathbf{T} \rangle$ *in* $\mathrm{L}_{\mathrm{q}}(\mathcal{K})$.

Note that the canonical join representation of $\langle \mathbf{T} \rangle$ can be found effectively given $\mathbf{T}$ and its $\mathcal{K}$-congruence lattice.

COROLLARY 2.16. *Every finite algebra is either quasicritical or a subdirect product of quasicritical subalgebras.*

The corollary has an elementary proof; see Lemma 4.1.

While we have shown that $\langle \mathbf{T} \rangle$ is completely join irreducible when $\mathbf{T}$ is quasicritical, we have not described a method for determining exactly which quasicritical algebras are in $\langle \mathbf{T} \rangle$. For any given algebra $\mathbf{A}$, one can determine whether $\mathbf{A} \leq \mathbf{T}^n$ for some n, by comparing the subdirect decompositions of $\mathbf{A}$ with a list of subalgebras of $\mathbf{T}$. (The Universal Algebra Calculator can be of assistance here.) But it is harder to tell whether, or when, a list of quasicritical algebras in $\langle \mathbf{T} \rangle$ is complete.

Indeed, a finitely generated quasivariety might contain uncountably many subquasivarieties. The classical example is that the quasivariety generated by the modular lattice $\mathbf{M}_{3,3}$ contains infinitely many quasicritical algebras and $2^{\aleph_0}$ subquasivarieties [82]. Lemma 5.26 and Corollary 5.27 show that the quasivariety $\langle \mathbf{T}_3 \rangle$ generated by a 3-element unary algebra likewise contains infinitely many quasicriticals and $2^{\aleph_0}$ subquasivarieties. Both these quasivarieties are in fact Q-universal; see Section 5.5.

2.2. Reflections

Given an algebra $\mathbf{T}$ in a quasivariety $\mathcal{K}$, and a subquasivariety $\mathcal{Q} \leq \mathcal{K}$, the *reflection* of $\mathcal{Q}$ in $\mathbf{T}$ is the congruence

$$\rho_{\mathcal{Q}}^{\mathbf{T}} = \bigwedge \{\theta \in \mathrm{Con}_{\mathcal{K}} \mathbf{T} : \mathbf{T}/\theta \in \mathcal{Q}\}.$$

When working with a single algebra, we may omit the superscript $\mathbf{T}$.

Note that if we fix a quasivariety $\mathcal{K}$, then for a subquasivariety $\mathcal{Q} \leq \mathcal{K}$ the reflection $\rho_{\mathcal{Q}}$ will be a $\mathcal{K}$-congruence, i.e., $\mathbf{T}/\rho_{\mathcal{Q}} \in \mathcal{K}$. As long as we are interested in the lattice of subquasivarieties $\mathrm{L}_{\mathrm{q}}(\mathcal{K})$, it makes sense to restrict our attention to the lattice $\mathrm{Con}_{\mathcal{K}} \mathbf{T}$ of all $\mathcal{K}$-congruences of $\mathbf{T}$, i.e., all congruences θ such that $\mathbf{T}/\theta \in \mathcal{K}$.

LEMMA 2.17. *Let* $\mathbf{T}$ *be a finite algebra in a quasivariety* $\mathcal{K}$ *of finite type. The reflection map* $\rho : \mathrm{L}_{\mathrm{q}}(\mathcal{K}) \to \mathrm{Con}_{\mathcal{K}} \mathbf{T}$ *has these properties.*

(1) ρ_Q is the least congruence such that $\mathbf{T}/\theta \in Q$.
(2) In particular, $\mathbf{T} \in Q$ if and only if $\rho_Q = \Delta$.
(3) If $Q \le R$, then $\rho_Q \ge \rho_R$.
(4) $\rho_{Q \vee R} = \rho_Q \wedge \rho_R$.
(5) More generally, $\rho_{\bigvee Q_i} = \bigwedge \rho_{Q_i}$.
(6) If $\rho_{Q_i} = \theta$ for all $i \in I$, then $\rho_{\bigwedge Q_i} = \theta$.

The proofs are left to the reader. For part (4), note that an algebra satisfies $\mathbf{S} \in Q \vee R$ if and only if $\mathbf{S}$ is a subdirect product of algebras in Q and R (Lemma 2.1). Then for (5), use compactness (Lemma 2.4).

Let us say that a congruence θ is a *reflection congruence* if $\theta = \rho_Q$ for some quasivariety Q. Note that reflection congruences are closed under meets, since $\bigwedge \rho_{Q_i} = \rho_{\bigvee Q_i}$. Thus the reflection congruences form a complete meet subsemilattice of $\mathrm{Con}_{\mathcal{K}} \mathbf{T}$. Of course, $\nabla = \rho_{\mathcal{T}}$ and $\Delta = \rho_{\mathcal{K}}$ are reflection congruences, where $\mathcal{T}$ denotes the trivial quasivariety generated by a 1-element structure with all possible relations of the type holding.

Because the reflection congruences on $\mathbf{T}$ are closed under arbitrary meets, they form a closure system. Thus we can define a *reflection closure operator* $\gamma^{\mathbf{T}}$ on $\mathrm{Con}_{\mathcal{K}} \mathbf{T}$. For $\theta \in \mathrm{Con}_{\mathcal{K}} \mathbf{T}$, let

$$\gamma^{\mathbf{T}}(\theta) = \bigwedge \{\rho_Q^{\mathbf{T}} : Q \in \mathrm{L}_q(\mathcal{K}) \text{ and } \rho_Q^{\mathbf{T}} \ge \theta\}.$$

Then $\gamma^{\mathbf{T}}$ is a closure operator, with $\gamma^{\mathbf{T}}(\theta)$ being the least reflection congruence on $\mathbf{T}$ containing θ. A $\mathcal{K}$-congruence θ is a reflection congruence if and only if $\gamma^{\mathbf{T}}(\theta) = \theta$. Again we omit the superscript $\mathbf{T}$ when the context is clear.

LEMMA 2.18. *Let $\theta \in \mathrm{Con}_{\mathcal{K}} \mathbf{T}$. Then*

(1) $\theta \ge \rho_{\langle \mathbf{T}/\theta \rangle}$;
(2) θ *is a reflection congruence if and only if $\theta = \rho_{\langle \mathbf{T}/\theta \rangle}$.*

PROOF. Let $\mathbf{B} = \mathbf{T}/\theta$ and $\mathcal{B} = \langle \mathbf{B} \rangle$. The first statement of the lemma follows from $\mathbf{B} = \mathbf{T}/\theta \in \mathcal{B}$. For the second part, if $\theta = \rho_{\mathcal{B}}$, then clearly it is a reflection congruence. Conversely, assume $\theta = \rho_Q$. Then $\mathbf{B} = \mathbf{T}/\theta \in Q$, whence $\mathcal{B} \le Q$ and $\rho_{\mathcal{B}} \ge \rho_Q$. Combining with (1) yields $\rho_{\mathcal{B}} \ge \rho_Q = \theta \ge \rho_{\mathcal{B}}$. $\qquad\square$

This gives us an effective method for deciding whether a given congruence θ is a reflection congruence.

THEOREM 2.19. *A congruence $\theta \in \mathrm{Con}_{\mathcal{K}} \mathbf{T}$ is a reflection congruence if and only if for all $\varphi \in \mathrm{Con}_{\mathcal{K}} \mathbf{T}$, $\mathbf{T}/\varphi \in \langle \mathbf{T}/\theta \rangle$ implies $\varphi \ge \theta$.*

The next result summarizes the options, followed by some useful consequences.

COROLLARY 2.20. *Let $\theta \in \mathrm{Con}_{\mathcal{K}} \mathbf{T}$.*

(1) *If θ is a reflection congruence, then $\rho_{\langle \mathbf{T}/\theta \rangle} = \theta = \gamma(\theta)$.*
(2) *If θ is not a reflection congruence, then $\rho_{\langle \mathbf{T}/\theta \rangle} < \theta < \gamma(\theta)$.*

COROLLARY 2.21. *Let α be an atom in* $\mathrm{Con}_{\mathcal{K}}\ \mathbf{T}$. *Then α is a reflection congruence if and only if* $\mathbf{T} \notin \langle \mathbf{T}/\alpha \rangle$.

COROLLARY 2.22. *If $\varphi \leq \psi \leq \theta$ in* $\mathrm{Con}_{\mathcal{K}}\ \mathbf{T}$ *and θ/ψ is not a reflection congruence in* $\mathbf{T}/\psi$, *then θ/φ is not a reflection congruence in* $\mathbf{T}/\varphi$.

Theorem 2.19 will be applied repeatedly in Chapter 5 to find the reflection congruences for small algebras.

An elementary argument shows that if a quotient algebra $\mathbf{T}/\theta$ satisfies a quasi-equation ε, and σ is an automorphism of $\mathbf{T}$, then $\mathbf{T}/\sigma(\theta)$ also satisfies ε. Hence $\mathbf{T}/\theta \in \Omega$ if and only if $\mathbf{T}/\sigma(\theta) \in \Omega$. This observation has the following useful consequence.

LEMMA 2.23. *Every reflection congruence is characteristic, that is,* $\sigma(\rho_\Omega) = \rho_\Omega$ *for every automorphism $\sigma \in \mathrm{Aut}\ \mathbf{T}$.*

Thus if $\theta \in \mathrm{Con}_{\mathcal{K}}\ \mathbf{T}$ and there is an automorphism σ such that $\sigma(\theta) \neq \theta$, then θ is not a reflection congruence. Indeed, whenever φ, θ are distinct congruences in $\mathrm{Con}_{\mathcal{K}}\ \mathbf{T}$ with $\mathbf{T}$ finite and $\mathbf{T}/\varphi \cong \mathbf{T}/\theta$, then of course $\mathbf{T}/\varphi \in \langle \mathbf{T}/\theta \rangle$, so $\rho_{\langle \mathbf{T}/\theta \rangle} \leq \varphi \wedge \theta < \theta$, and θ is not a reflection congruence. In practice, this comes up often.

The results thus far in this section did not require local finiteness. At this point in the discussion, let us fix a locally finite quasivariety $\mathcal{K}$ of finite type. Our long-range goals are to describe completely meet irreducible quasi-varieties in $\mathrm{L_q}(\mathcal{K})$ (Theorem 2.30) and to find $\kappa(\mathbf{T})$ for quasicritical algebras (Theorem 2.32). The analysis requires some setup.

Let $\mathbf{T}$ be a finite (not necessarily quasicritical) algebra in $\mathcal{K}$. Let X be a finite set of variables, large enough so that $\mathbf{T}$ has a generating set with $|X|$ elements. Let $\mathbf{W}(X)$ be the term algebra (absolutely free algebra) in the type of $\mathcal{K}$ generated by X, and let $\mathbf{F}_{\mathcal{K}}(X)$ be the free algebra over $\mathcal{K}$ generated by X. Set $\varphi : \mathbf{W}(X) \twoheadrightarrow \mathbf{F}_{\mathcal{K}}(X)$ to be the homomorphism that evaluates each term in the free algebra. Finally, for a pair of terms u, $v \in \mathbf{W}(X)$, let $\mathrm{Cg}_{\mathcal{K}}(u \approx v)$ denote the $\mathcal{K}$-congruence on $\mathbf{F}_{\mathcal{K}}(X)$ generated by $(\varphi(u), \varphi(v))$.

Fix a surjective homomorphism $m : \mathbf{F}_{\mathcal{K}}(X) \twoheadrightarrow \mathbf{T}$. For each element a in T, pick a term $t_a \in W(X)$ in $(m\varphi)^{-1}(a)$. Set H to be the set of equations

$$H = \{ f(t_{a_1}, \dots, t_{a_n}) \approx t_b \mid f \text{ is an } n\text{-ary basic operation and}$$
$$f^{\mathbf{T}}(a_1, \dots, a_n) = b \}.$$

The *diagram* of $\mathbf{T}$ is the conjunction

$$\underset{\psi \in H}{\&}\ \psi.$$

The point of all this setup is that we have found pairs of terms so that

$$\bigvee_{\psi \in H} \mathrm{Cg}_{\mathcal{K}}(\psi) = \ker m$$

in $\mathrm{Con}_{\mathcal{K}}\ \mathbf{F}_{\mathcal{K}}(X)$.

Often the conjunction H includes more information than necessary. In practice, it is convenient to use $\mathrm{diag}(\mathbf{T})$ to denote any conjunction of equations $\psi_1 \,\&\, \psi_2 \,\&\, \cdots \,\&\, \psi_k$ with each $\psi_i \in H$ such that

$$\mathrm{Cg}_{\mathcal{K}}(\psi_1) \vee \cdots \vee \mathrm{Cg}_{\mathcal{K}}(\psi_k) = \ker m.$$

For each minimal congruence $\alpha \succ \Delta$ in $\mathrm{Con}_{\mathcal{K}} \mathbf{T}$, choose elements $a, b \in T$ such that $\alpha = \mathrm{Cg}_{\mathcal{K}}(a, b)$. Then form the quasi-equation

$$\varepsilon_{\mathbf{T},\alpha} : \quad \mathrm{diag}(\mathbf{T}) \to t_a \approx t_b.$$

For example, if $\mathbf{T}$ is the unary algebra $(\{a, b, c\}, f)$ with $f(a) = b$ and $f(b) = c = f(c)$, and $\alpha = \mathrm{Cg}(b, c)$, then $\varepsilon_{\mathbf{T},\alpha}$ can be written as

$$f^3(x) \approx f^2(x) \to f^2(x) \approx f(x).$$

THEOREM 2.24. *Let $\mathcal{K}$ be a locally finite quasivariety of finite type, and let $\mathbf{T}$ be a finite algebra in $\mathcal{K}$. The following are equivalent for a quasivariety $\mathcal{Q} \leq \mathcal{K}$ and an atom α in $\mathrm{Con}_{\mathcal{K}} \mathbf{T}$.*

(1) $\mathcal{Q} \leq \langle \varepsilon_{\mathbf{T},\alpha} \rangle$ *in* $\mathrm{L_q}(\mathcal{K})$.
(2) $\varepsilon_{\mathbf{T},\alpha}$ *is in the theory of* $\mathcal{Q}$.
(3) *For every $\mathbf{A} \in \mathcal{Q}$, $\mathbf{A}$ satisfies $\varepsilon_{\mathbf{T},\alpha}$.*
(4) *For every $\mathbf{A} \in \mathcal{Q}$, if $h : \mathbf{T} \to \mathbf{A}$ is a homomorphism, then $\alpha \leq \ker h$.*
(5) $\alpha \leq \rho_{\mathcal{Q}}^{\mathbf{T}}$.
(6) $\gamma(\alpha) \leq \rho_{\mathcal{Q}}^{\mathbf{T}}$.

Thus $\mathcal{Q}$ satisfies $\varepsilon_{\mathbf{T},\alpha}$ if and only if $\alpha \leq \rho_{\mathcal{Q}}^{\mathbf{T}}$.

Indeed, it is easy to see that each successive pair is equivalent, and this yields in particular (2) $\Leftrightarrow$ (5). In words, Theorem 2.24 says that $\langle \varepsilon_{\mathbf{T},\alpha} \rangle$ *is the largest quasivariety $\mathcal{Q}$ such that $\rho_{\mathcal{Q}}^{\mathbf{T}} \geq \alpha$.*

COROLLARY 2.25. *If α is an atom in $\mathrm{Con}_{\mathcal{K}} \mathbf{T}$, then $\gamma(\alpha) = \rho_{\langle \varepsilon_{\mathbf{T},\alpha} \rangle}^{\mathbf{T}}$.*

PROOF. First, take $\mathcal{Q} = \langle \varepsilon_{\mathbf{T},\alpha} \rangle$ in Theorem 2.24. Then (1) $\Rightarrow$ (6) yields $\gamma(\alpha) \leq \rho_{\langle \varepsilon_{\mathbf{T},\alpha} \rangle}^{\mathbf{T}}$. For the reverse inclusion, choose a quasivariety $\mathcal{R}$ such that $\gamma(\alpha) = \rho_{\mathcal{R}}^{\mathbf{T}}$. Then (6) $\Rightarrow$ (1) yields $\mathcal{R} \leq \langle \varepsilon_{\mathbf{T},\alpha} \rangle$. Taking reflections, we obtain $\gamma(\alpha) = \rho_{\mathcal{R}}^{\mathbf{T}} \geq \rho_{\langle \varepsilon_{\mathbf{T},\alpha} \rangle}^{\mathbf{T}}$, as desired. □

Note that the equivalence of (2) and (4) in Theorem 2.24 means that, for practical purposes, $\varepsilon_{\mathbf{T},\alpha}$ is independent of the choice of the pair (a, b) generating α. For if $\alpha = \mathrm{Cg}_{\mathcal{K}}(a, b) = \mathrm{Cg}_{\mathcal{K}}(c, d)$, then the quasi-equation $\mathrm{diag}(\mathbf{T}) \to t_a \approx t_b$ is equivalent to $\mathrm{diag}(\mathbf{T}) \to t_c \approx t_d$.

THEOREM 2.26. *Let $\mathcal{K}$ be a locally finite quasivariety of finite type. Let $\mathbf{T}$ be a finite algebra in $\mathcal{K}$, and let $\alpha_1, \ldots, \alpha_n$ be the set of atoms in $\mathrm{Con}_{\mathcal{K}} \mathbf{T}$. Let $\mathcal{Q}$ be any quasivariety in $\mathrm{L_q}(\mathcal{K})$. Then $\mathbf{T} \notin \mathcal{Q}$ if and only if $\mathcal{Q}$ satisfies $\varepsilon_{\mathbf{T},\alpha_i}$ for some i.*

PROOF. Clearly, if the quasivariety $\mathfrak{Q}$ satisfies $\varepsilon_{\mathbf{T},\alpha}$ for some atom α, then $\mathbf{T} \notin \mathfrak{Q}$ because $\mathbf{T}$ fails that quasi-equation under the substitution by $m\varphi$, where $m\varphi : \mathbf{W}(X) \to \mathbf{T}$ is the map used to construct $\varepsilon_{\mathbf{T},\alpha}$.

Conversely, assume $\mathbf{T} \notin \mathfrak{Q}$. Then $\rho_{\mathfrak{Q}}^{\mathbf{T}} > \Delta$, and hence $\rho_{\mathfrak{Q}}^{\mathbf{T}} \geq \alpha$ for at least one atom α of $\mathrm{Con}_{\mathcal{K}}\, \mathbf{T}$. By Lemma 2.24, this implies that $\mathfrak{Q}$ satisfies $\varepsilon_{\mathbf{T},\alpha}$ for this α. $\qquad\square$

COROLLARY 2.27. *If $\mathbf{T}$ is a finite algebra in a locally finite quasivariety $\mathcal{K}$ of finite type, then the quasivariety $\langle \mathbf{T} \rangle$ has finitely many lower covers in $L_q(\mathcal{K})$, each of the form $\langle \mathbf{T} \rangle \wedge \langle \varepsilon_{\mathbf{T},\alpha} \rangle$ for some atom α of $\mathrm{Con}_{\mathcal{K}}\, \mathbf{T}$.*

Clearly, the quasi-equation $\varepsilon_{\mathbf{T},\alpha}$ can be defined in the same way for any principal congruence in $\mathrm{Con}_{\mathcal{K}}\, \mathbf{T}$. Theorem 2.26 shows why the quasi-equations for atoms of the $\mathcal{K}$-congruence lattice are our primary interest.

If a finite algebra $\mathbf{T}$ is quasicritical, then the proper subalgebras of $\mathbf{T}$ generate a quasivariety $\mathcal{S}$ with $\mathbf{T} \notin \mathcal{S}$. Then $\rho_{\mathcal{S}}^{\mathbf{T}} \geq \alpha$ for some atom α of $\mathrm{Con}_{\mathcal{K}}\, \mathbf{T}$, and hence every proper subalgebra of $\mathbf{T}$ satisfies the corresponding quasi-equation $\varepsilon_{\mathbf{T},\alpha}$.

On the other hand, $\langle \mathbf{T} \rangle$ is join irreducible in $L_q(\mathcal{K})$ when $\mathbf{T}$ is quasicritical, so $\mathcal{S} \leq \langle \mathbf{T} \rangle_*$, while $\langle \mathbf{T} \rangle_* = \langle \mathbf{T} \rangle \wedge \langle \varepsilon_{\mathbf{T},\beta} \rangle$ for some (possibly different) atom β. There is a useful criterion to decide whether the lower cover $\langle \mathbf{T} \rangle_*$ is the quasivariety $\mathcal{S}$ generated by the proper subalgebras of $\mathbf{T}$.

THEOREM 2.28. *Let $\mathbf{T}$ be a quasicritical algebra in a locally finite quasivariety $\mathcal{K}$ of finite type, and let $\mathcal{S}$ be the quasivariety generated by the proper subalgebras of $\mathbf{T}$. The following are equivalent.*

(1) *For all $n \geq 2$, if $\mathbf{U} \leq \mathbf{T}^n$, then either $\mathbf{U} \in \mathcal{S}$ or $\mathbf{T} \leq \mathbf{U}$.*
(2) $\mathcal{S} = \langle \mathbf{T} \rangle_*$.

PROOF. First assume (1), and consider any finite algebra $\mathbf{U} \in \langle \mathbf{T} \rangle$. Then $\mathbf{U} \leq \mathbf{T}^n$ for some n, whence by (1) either $\mathbf{U} \in \mathcal{S}$ or $\mathbf{T} \leq \mathbf{U}$. In the latter case, $\langle \mathbf{U} \rangle = \langle \mathbf{T} \rangle$. Thus every finite algebra in $\langle \mathbf{T} \rangle$ that does not generate $\langle \mathbf{T} \rangle$ is contained in $\mathcal{S}$. Since every subquasivariety of $\mathcal{K}$ is determined by the finite algebras it contains (by local finiteness), we conclude that $\mathcal{S} = \langle \mathbf{T} \rangle_*$.

Conversely, assuming that $\mathcal{S} = \langle \mathbf{T} \rangle_*$, consider a subalgebra $\mathbf{U} \leq \mathbf{T}^n$. If $\langle \mathbf{U} \rangle < \langle \mathbf{T} \rangle$, then $\mathbf{U} \in \langle \mathbf{T} \rangle_* = \mathcal{S}$. On the other hand, if $\langle \mathbf{U} \rangle = \langle \mathbf{T} \rangle$, then $\mathbf{T} \leq \mathbf{U}$ by Lemma 2.9 with $m = 1$. Thus (1) holds. $\qquad\square$

Problem 2 of Chapter 8 asks whether there is a bound B, perhaps depending on $\mathbf{T}$, such that it suffices to check property (1) for $n \leq B$. Meanwhile, the algebra seminar at Latrobe University (Brian Davey, Lucy Ham, Marcel Jackson, and Tomasz Kowalski) observed that a slightly stronger condition can often be used to show that $\mathcal{S} = \langle \mathbf{T} \rangle_*$.

COROLLARY 2.29. *Let $\mathbf{T}$ be a quasicritical algebra in a locally finite quasivariety $\mathcal{K}$ of finite type, and let $\mathcal{S}$ be the quasivariety generated by the proper subalgebras of $\mathbf{T}$. The following are equivalent.*

(1) *If $\mathbf{U} \leq \mathbf{T}^2$ and either coordinate projection is onto $\mathbf{T}$, then $\mathbf{T} \leq \mathbf{U}$.*

(2) *For all $n \geq 2$, if $\mathbf{U} \leq \mathbf{T}^n$ and some coordinate projection is onto $\mathbf{T}$, then $\mathbf{T} \leq \mathbf{U}$.*

Moreover, these conditions imply that $\mathcal{S} = \langle \mathbf{T} \rangle_*$.

The proof of equivalence, by induction on n, is left to the reader. Clearly property (1) of the corollary is easily checked, while property (2) implies Theorem 2.28(1).

2.3. Completely Meet Irreducible Quasivarieties

Now let us apply the results of the previous section to the structure of $\mathrm{L}_q(\mathcal{K})$. Recall that every quasivariety in $\mathrm{L}_q(\mathcal{K})$ is a meet of completely meet irreducible quasivarieties. Our next result identifies those quasivarieties that are completely meet irreducible.

THEOREM 2.30. *Let $\mathcal{K}$ be a locally finite quasivariety of finite type. Let $\mathcal{Q}$ be a completely meet irreducible quasivariety in $\mathrm{L}_q(\mathcal{K})$. Then $\mathcal{Q} = \langle \varepsilon_{\mathbf{T},\alpha} \rangle$ for a unique quasicritical algebra $\mathbf{T} \in \mathcal{K}$ and some congruence $\alpha \succ \Delta$ in $\mathrm{Con}_{\mathcal{K}} \mathbf{T}$.*

PROOF. Let $\mathcal{Q}$ be a completely meet irreducible quasivariety, and let $\mathcal{Q}^*$ denote its unique upper cover in $\mathrm{L}_q(\mathcal{K})$. Then $\mathcal{Q} = \langle \varphi \rangle$ for any quasi-equation φ that is in the theory of $\mathcal{Q}$ but not in the theory of $\mathcal{Q}^*$.

Choose any quasicritical algebra $\mathbf{T} \in \mathcal{Q}^* \setminus \mathcal{Q}$, and apply Theorem 2.26. Then $\mathbf{T} \in \mathcal{Q}^*$ implies that $\mathcal{Q}^*$ fails $\varepsilon_{\mathbf{T},\alpha}$ for every atom α of $\mathrm{Con}_{\mathcal{K}} \mathbf{T}$. However, $\mathbf{T} \notin \mathcal{Q}$ implies that $\mathcal{Q}$ satisfies $\varepsilon_{\mathbf{T},\alpha_0}$ for some $\alpha_0 \succ \Delta$. By the preceding paragraph, we have $\mathcal{Q} = \langle \varepsilon_{\mathbf{T},\alpha_0} \rangle$.

Note that $\mathcal{Q}^* = \mathcal{Q} \vee \langle \mathbf{T} \rangle$ in $\mathrm{L}_q(\mathcal{K})$. By the join semidistributivity of $\mathrm{L}_q(\mathcal{K})$, this makes $\langle \mathbf{T} \rangle$ unique, whence by Theorem 2.11 there is a unique choice for $\mathbf{T}$. □

The converse of Theorem 2.30 is false. Examples of meet reducible quasivarieties $\langle \varepsilon_{\mathbf{T},\alpha} \rangle$ with $\mathbf{T}$ quasicritical and $\alpha \succ \Delta$ are given in Propositions 5.25(11), 5.29 and 7.9. The situation when this occurs is described in Corollary 3.13.

Recall that for a finite algebra $\mathbf{T}$ in $\mathcal{K}$, $\kappa(\mathbf{T})$ denotes the set of all quasivarieties $\mathcal{N} \leq \mathcal{K}$ that are maximal with respect to the property that $\mathbf{T} \notin \mathcal{N}$, and that any $\mathcal{N} \in \kappa(\mathbf{T})$ is completely meet irreducible. The rest of this section is devoted to results for deciding which quasivarieties $\langle \varepsilon_{\mathbf{T},\alpha} \rangle$ comprise $\kappa(\mathbf{T})$. The quasivarieties $\langle \varepsilon_{\mathbf{T},\alpha} \rangle$ for different atoms α need not all be distinct, nor need they all be in $\kappa(\mathbf{T})$.

THEOREM 2.31. *For atoms α, $\beta \succ \Delta$ in $\mathrm{Con}_{\mathcal{K}} \mathbf{T}$, the following are equivalent.*

(1) $\varepsilon_{\mathbf{T},\beta}$ *implies* $\varepsilon_{\mathbf{T},\alpha}$.
(2) $\langle \varepsilon_{\mathbf{T},\beta} \rangle \leq \langle \varepsilon_{\mathbf{T},\alpha} \rangle$ *in* $\mathrm{L}_q(\mathcal{K})$.
(3) $\rho^{\mathbf{T}}_{\langle \varepsilon_{\mathbf{T},\beta} \rangle} \geq \rho^{\mathbf{T}}_{\langle \varepsilon_{\mathbf{T},\alpha} \rangle}$.
(4) $\gamma(\beta) \geq \gamma(\alpha)$ *in* $\mathrm{Con}_{\mathcal{K}} \mathbf{T}$.

PROOF. Clearly, (1) and (2) are equivalent, while (3) and (4) are equivalent per Corollary 2.25. Moreover, (2) implies (3) because reflection is order-reversing. Conversely, assume that (3) holds, so that $\rho^{\mathbf{T}}_{\langle \varepsilon_{\mathbf{T},\beta}\rangle} \geq \rho^{\mathbf{T}}_{\langle \varepsilon_{\mathbf{T},\alpha}\rangle} = \gamma(\alpha)$. Applying the implication (6) $\Rightarrow$ (2) of Theorem 2.24 to α and $\mathfrak{Q} = \langle \varepsilon_{\mathbf{T},\beta}\rangle$ yields $\langle \varepsilon_{\mathbf{T},\beta}\rangle \leq \langle \varepsilon_{\mathbf{T},\alpha}\rangle$, which is (2). □

Because reflection reverses order, we have the following consequence.

THEOREM 2.32. $\langle \varepsilon_{\mathbf{T},\alpha}\rangle \in \kappa(\mathbf{T})$ *if and only if* $\gamma(\alpha)$ *is minimal among the nonzero reflection congruences of* $\mathbf{T}$.

We should be very explicit about this connection. Given a completely meet irreducible quasivariety $\mathfrak{Q}$, there is a unique finite quasicritical algebra $\mathbf{T}$ such that $\langle \mathbf{T}\rangle$ is the unique minimal quasivariety below $\mathfrak{Q}^*$ but not below $\mathfrak{Q}$, so that $\mathfrak{Q} \in \kappa(\mathbf{T})$. Then $\mathfrak{Q} = \langle \varepsilon_{\mathbf{T},\alpha}\rangle$ for some atom $\alpha \succ \Delta$ in $\mathrm{Con}_{\mathcal{K}}\,\mathbf{T}$. Moreover, $\gamma(\alpha)$ is a minimal nonzero reflection congruence in $\mathrm{Con}_{\mathcal{K}}\,\mathbf{T}$. However, there may be other minimal reflection congruences $\gamma(\beta)$ for $\mathbf{T}$. For each such minimal reflection congruence, $\langle \mathbf{T}\rangle$ is the unique minimal congruence below $\langle \varepsilon_{\mathbf{T},\beta}\rangle^*$ but not below $\langle \varepsilon_{\mathbf{T},\beta}\rangle$, and thus $\langle \varepsilon_{\mathbf{T},\beta}\rangle \in \kappa(\mathbf{T})$. This distinction reflects the fact that $\mathrm{L}_q(\mathcal{K})$ is join semidistributive, but not necessarily meet semidistributive.

However, $\langle \mathbf{T}\rangle$ need not be the only quasivariety minimal with respect to being not below $\langle \varepsilon_{\mathbf{T},\alpha}\rangle$ when the latter is completely meet irreducible. There may be other quasivarieties with this property. These are each generated by a finite quasicritical algebra, but by join semidistributivity $\langle \mathbf{T}\rangle$ will be the only one that is below $\langle \varepsilon_{\mathbf{T},\alpha}\rangle^*$. The set of algebras $\mathbf{R}$ such that $\langle \mathbf{R}\rangle$ is minimal with respect to not being below $\langle \varepsilon_{\mathbf{T},\alpha}\rangle$ will be denoted $\mathbf{N}(\mathbf{T}, \alpha)$. We give an algorithm for finding it in the next section.

The generic situation surrounding a completely meet irreducible quasivariety $\mathfrak{Q}$ in $\mathrm{L}_q(\mathcal{K})$ is illustrated in Figure 2.1. The corresponding diagram for a completely join irreducible quasivariety $\langle \mathbf{T}\rangle$ is given in Figure 2.2.

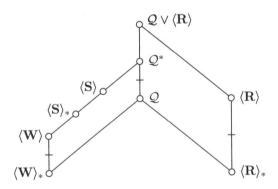

FIGURE 2.1. Typical situation starting with $\mathfrak{Q} = \langle \varepsilon_{\mathbf{T},\alpha}\rangle$ completely meet irreducible. Here $\langle \mathbf{T}\rangle_* \leq \mathfrak{Q}$ and $\langle \mathbf{T}\rangle \leq \mathfrak{Q}^*$, whence $\mathfrak{Q} \in \kappa(\mathbf{T})$, while $\langle \mathbf{R}\rangle_* \leq \mathfrak{Q}$ but $\langle \mathbf{R}\rangle \nleq \mathfrak{Q}^*$, so that $\langle \mathbf{R}\rangle \in \mathbf{N}(\mathbf{T}, \alpha)$ although $\mathfrak{Q} \notin \kappa(\mathbf{R})$. Ticks indicate covers.

COROLLARY 2.33. *The following are equivalent for a finite algebra* $\mathbf{T} \in \mathcal{K}$.

(1) $\langle \mathbf{T} \rangle$ *is join prime in* $\mathrm{L_q}(\mathcal{K})$.
(2) $\langle \mathbf{T} \rangle$ *is completely join prime in* $\mathrm{L_q}(\mathcal{K})$.
(3) $|\kappa(\mathbf{T})| = 1$.
(4) *There is a unique minimal nonzero reflection congruence in* $\mathrm{Con}_{\mathcal{K}} \mathbf{T}$.

Moreover, if $\mathbf{T}$ *satisfies these properties and* $\alpha \in \mathrm{Con}_{\mathcal{K}} \mathbf{T}$ *is an atom such that* $\gamma(\alpha)$ *is the minimal nonzero reflection congruence, then for every sub-quasivariety* $\mathcal{Q} \leq \mathcal{K}$, *either* $\langle \mathbf{T} \rangle \leq \mathcal{Q}$ *or* $\mathcal{Q} \leq \langle \varepsilon_{\mathbf{T},\alpha} \rangle$. *Thus* $\langle \varepsilon_{\mathbf{T},\alpha} \rangle$ *is completely meet prime in* $\mathrm{L_q}(\mathcal{K})$.

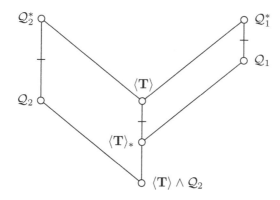

FIGURE 2.2. Typical situation starting with $\langle \mathbf{T} \rangle$ completely join irreducible. Here $\mathcal{Q}_1 = \langle \varepsilon_{\mathbf{T},\alpha} \rangle$ and $\mathcal{Q}_2 = \langle \varepsilon_{\mathbf{T},\beta} \rangle$, say. It can also happen that $\langle \mathbf{T} \rangle \wedge \mathcal{Q}_2 = \langle \mathbf{T} \rangle_*$. Ticks indicate covers.

Corollaries 3.5 and 3.8 contain additional information along these lines. In particular, the corollary above applies when $\mathrm{Con}_{\mathcal{K}} \mathbf{T}$ has a unique atom, i.e., when $\mathbf{T}$ is $\mathcal{K}$-subdirectly irreducible.

Let us introduce one more bit of terminology. McKenzie showed in [115] that a finite, subdirectly irreducible lattice $\mathbf{L}$ that is a bounded homomorphic image of a free lattice has a *splitting equation* φ with the property that for any lattice variety $\mathcal{V}$, we have $\mathbf{L} \notin \mathcal{V}$ if and only if $\mathcal{V}$ satisfies φ. For example, the splitting equation for the pentagon $\mathbf{N}_5$ in lattices is the modular law. Analogously, for any finite structure $\mathbf{T}$ in a locally finite quasivariety $\mathcal{K}$ of finite type, we have that $\mathbf{T} \notin \mathcal{Q}$ for a subquasivariety $\mathcal{Q} \leq \mathcal{K}$ if and only if $\mathcal{Q}$ satisfies $\varepsilon_{\mathbf{T},\alpha}$ for some atom α of $\mathrm{Con}_{\mathcal{K}} \mathbf{T}$ (Theorem 2.26). Hence we refer to the quasi-equations $\varepsilon_{\mathbf{T},\alpha}$ as the *semi-splitting quasi-equations* of $\mathbf{T}$; their satisfaction by a quasivariety corresponds to omitting the structures in

$N(\mathbf{T}, \alpha)$ (Theorem 3.4). If $\mathbf{T}$ has a unique minimal reflection congruence, so that there is up to equivalence only one $\varepsilon_{\mathbf{T}, \alpha}$, then we refer to $\varepsilon_{\mathbf{T}, \alpha}$ as the *splitting quasi-equation* for $\mathbf{T}$.

Remark. McKenzie's characterization of splitting lattices, i.e., lattices that generate a completely join prime lattice variety, is that $\mathbf{L}$ *is a splitting lattice if and only if* $\mathbf{L}$ *is finite, subdirectly irreducible, and a bounded homomorphic image of a free lattice* [115]. For other varieties of algebras, we have the following. Recall that an algebra $\mathbf{A}$ is *weakly projective* in a class $\mathcal{K}$ if $\mathbf{A} \in \mathcal{K}$ and whenever $\mathbf{A}$ is a homomorphic image of an algebra $\mathbf{B} \in \mathcal{K}$, then $\mathbf{A}$ embeds into $\mathbf{B}$. (This is slightly weaker than the categorical notion of *projective*.) It is easy to see that a finite, weakly projective, subdirectly irreducible algebra $\mathbf{A}$ in a variety $\mathcal{V}$ generates a subvariety $\mathbb{V}(\mathbf{A})$ that is completely join prime in $L_v(\mathcal{V})$. In that case, $\mathbf{A}$ has a splitting equation φ with the property that for any subvariety $\mathcal{W} \leq \mathcal{V}$, we have $\mathbf{A} \notin \mathcal{W}$ if and only if $\mathcal{W}$ satisfies φ. Splitting lattices need not be weakly projective, so this condition is sufficient but not necessary for an algebra to generate a splitting variety.

In Chapters 5–7 there are many examples where $\langle \varepsilon_{\mathbf{T}, \alpha} \rangle$ is not in $\kappa(\mathbf{T})$ for one or more atoms α. A sufficient condition is provided by the next corollary, which is a consequence of Corollary 2.21 and Theorem 2.32.

COROLLARY 2.34. *If α is an atom in* $\mathrm{Con}_{\mathcal{K}}\, \mathbf{T}$ *and* $\mathbf{T}/\alpha$ *satisfies* $\varepsilon_{\mathbf{T}, \alpha}$, *then α is a reflection congruence and* $\langle \varepsilon_{\mathbf{T}, \alpha} \rangle \in \kappa(\mathbf{T})$.

2.4. Quasivarieties of Modular Lattices

In this section we examine some quasivarieties of modular lattices.

One of the first real theorems of lattice theory is due to Richard Dedekind [50].

THEOREM 2.35. *The following are equivalent for a lattice* $\mathbf{L}$.

(1) $\mathbf{L}$ *satisfies the quasi-equation*
$$x \geq y \rightarrow x \wedge (y \vee z) \approx y \vee (x \wedge z).$$

(2) $\mathbf{L}$ *satisfies the equation*
$$x \wedge ((x \wedge y) \vee z) \approx (x \wedge y) \vee (x \wedge z).$$

(3) $\mathbf{L}$ *does not contain the pentagon* $\mathbf{N}_5$ *as a sublattice.*

A lattice satisfying the properties of Theorem 2.35 is called *modular*. Let $\mathfrak{M}$ denote the class of all modular lattices. Dedekind's characterization of modular lattices by exclusion of the pentagon as a sublattice was the motivation for McKenzie's description of splitting lattices and splitting varieties [115].

Dedekind also showed that if a modular lattice $\mathbf{L}$ has a finite maximal chain of length k, then every maximal chain in $\mathbf{L}$ has length k. We then say that k is the *dimension* of $\mathbf{L}$. Recall that every finite dimensional, subdirectly irreducible, modular lattice is simple [32].

Our goal is to describe a lower segment of the subquasivariety lattice $L_q(\mathfrak{M})$. The results are illustrated in Figure 2.5. In this section, because we are dealing with both varieties and quasivarieties, we use the notation $\mathbb{V}(\mathfrak{X})$ and $\mathbb{Q}(\mathfrak{X})$. Thus $\mathbb{V}(\mathfrak{X}) = \mathbb{HSP}(\mathfrak{X})$ and $\mathbb{Q}(\mathfrak{X}) = \mathbb{SPU}(\mathfrak{X}) = \langle \mathfrak{X} \rangle$.

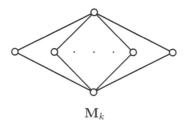

$\mathbf{M}_k$

FIGURE 2.3. The lattice $\mathbf{M}_k$ with k atoms.

Jónsson's Lemma, here stated for lattices, is a basic tool for congruence distributive varieties [91].

LEMMA 2.36. *Let $\mathfrak{X}$ be a collection of lattices. The finitely subdirectly irreducible lattices in $\mathbb{V}(\mathfrak{X})$ are contained in $\mathbb{HSU}(\mathfrak{X})$.*

For $k \geq 3$, including infinite cardinals, let $\mathbf{M}_k$ be the 2-dimensional lattice with k atoms; see Figure 2.3. In particular, $\mathbf{M}_\omega$ has a countably infinite number of atoms. It turns out that the variety and quasivariety generated by $\mathbf{M}_\omega$ are the same.

THEOREM 2.37. $\mathbb{V}(\mathbf{M}_\omega) = \mathbb{Q}(\mathbf{M}_\omega)$.

Theorem 2.37 is a special case of a slightly more general result. For an integer $m \geq 0$, let $\mathcal{L}_m$ be the class of lattices $\mathbf{K}$ such that every maximal chain of $\mathbf{K}$ has length at most m. For such a lattice, the *length* $\ell(\mathbf{K})$ is the length of the longest chain in $\mathbf{K}$. (The term *dimension* is normally reserved for lattices in which every maximal chain has the same length, such as modular or semimodular lattices.)

THEOREM 2.38. *For each $m \geq 0$, $\mathbb{V}(\mathcal{L}_m) = \mathbb{Q}(\mathcal{L}_m)$.*

PROOF. As always, $\mathbb{Q}(\mathcal{L}_m) \subseteq \mathbb{V}(\mathcal{L}_m)$. For the reverse inclusion, recall that the subdirectly irreducible lattices in $\mathbb{V}(\mathcal{L}_m)$ are contained in $\mathbb{HSU}(\mathcal{L}_m)$ by Jónsson's Lemma. Since having the length of chains bounded by m is a first-order property, the class $\mathcal{L}_m$ is closed under ultraproducts. Even more clearly, $\mathcal{L}_m$ is closed under sublattices and homomorphisms. Thus each subdirectly irreducible lattice in $\mathbb{V}(\mathcal{L}_m)$ is itself in $\mathcal{L}_m$. Every lattice in $\mathbb{V}(\mathcal{L}_m)$ is a subdirect product of these, whence $\mathbb{V}(\mathcal{L}_m) \subseteq \mathbb{Q}(\mathcal{L}_m)$. □

Theorem 2.37 is just the case $m = 2$ of Theorem 2.38. Note that for any infinite cardinal κ, $\mathbb{V}(\mathbf{M}_\kappa) = \mathbb{V}(\mathbf{M}_\omega) = \mathbb{Q}(\mathbf{M}_\omega) = \mathbb{Q}(\mathbf{M}_\kappa)$, since varieties and quasivarieties are determined by their finitely generated members.

We also use the following variant.

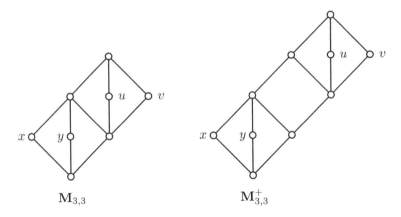

FIGURE 2.4. The lattices $\mathbf{M}_{3,3}$ and $\mathbf{M}_{3,3}^+$

THEOREM 2.39. *Let* $\mathbf{L}$ *be a finite lattice with the property that every subdirectly irreducible lattice in* $\mathbb{HS}(\mathbf{L})$ *is isomorphic to a sublattice of* $\mathbf{L}$. *Then* $\mathbb{V}(\mathbf{L}) = \mathbb{Q}(\mathbf{L})$.

PROOF. By Jónsson's Lemma, the hypothesis implies that every subdirectly irreducible lattice in $\mathbb{V}(\mathbf{L})$ is isomorphic to a sublattice of $\mathbf{L}$, whence $\mathbb{V}(\mathbf{L}) \subseteq \mathbb{Q}(\mathbf{L})$. $\square$

Theorem 2.39 applies to the lattices $\mathbf{M}_k$ for $3 \le k < \omega$ and the first lattice $\mathbf{M}_{3,3}$ in Figure 2.4. Likewise, the theorem applies to the lattices $\mathbf{M}_{k,\ell}$ with k atoms in the lower diamond and ℓ atoms in the upper one, where $k, \ell \ge 3$.

Now consider the second lattice $\mathbf{M}_{3,3}^+$ in Figure 2.4. Note that $\mathbf{M}_{3,3}^+$ is a quasicritical lattice that is a subdirect product of $\mathbf{M}_{3,3}$ and $\mathbf{2}$. Thus $\mathbb{Q}(\mathbf{M}_{3,3}^+) < \mathbb{Q}(\mathbf{M}_{3,3})$ while $\mathbb{V}(\mathbf{M}_{3,3}^+) = \mathbb{V}(\mathbf{M}_{3,3})$.

The main result of this section extends to quasivarieties a classic result of Jónsson for varieties of modular lattices. Jónsson's result [92] is the equivalence of parts (1), (2) of the next theorem and $\mathbf{M}_{3,3} \notin \mathbb{V}(\mathbf{L})$. (An earlier version is in Grätzer [80].)

THEOREM 2.40. *The following are equivalent for a modular lattice* $\mathbf{L}$.
(1) $\mathbf{L} \in \mathbb{V}(\mathbf{M}_\omega)$.
(2) $\mathbf{L}$ *satisfies the equation*
$$(\xi_2) \quad x \wedge (y \vee (u \wedge v)) \wedge (u \vee v) \le y \vee (x \wedge u) \vee (x \wedge v).$$
(3) $\mathbf{L}$ *satisfies the equation*
$$(\xi_3) \quad x \wedge (y \vee (u \wedge v)) \wedge (u \vee v) \le u \vee (v \wedge (x \vee y)) \vee (x \wedge y).$$
(4) $\mathbf{M}_{3,3}^+ \notin \mathbb{Q}(\mathbf{L})$.
(5) $\mathbf{L}$ *satisfies the quasi-equation*
$$(\xi_5) \quad \mathrm{diag}(\mathbf{M}_{3,3}^+) \to x \approx y.$$

Figure 2.5 illustrates the theorem.

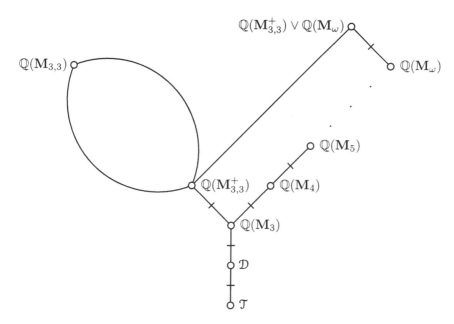

FIGURE 2.5. Segment of the lattice $L_q(\mathfrak{M})$ of quasivarieties of modular lattices. Here $\mathfrak{I}$ denotes all 1-element lattices, $\mathcal{D}$ denotes distributive lattices. Ticks indicate covers. The interval $[\mathbb{Q}(\mathbf{M}_{3,3}^+), \mathbb{Q}(\mathbf{M}_{3,3})]$ contains uncountably many quasivarieties.

PROOF. The equivalence of (1) and (2) is due to Jónsson [92]. The lattice $\mathbf{M}_\omega$ satisfies (ξ_3), so (1) implies (3), while $\mathbf{M}_{3,3}$ fails (ξ_3), whence (3) implies (4).

The congruence lattice $\mathrm{Con}\,\mathbf{M}_{3,3}^+$ is isomorphic to $\mathbf{2} \times \mathbf{2}$. One atom, $\alpha = \mathrm{Cg}(z, z')$ using the labels in Figure 2.6, is not a reflection congruence by Lemma 2.18(2), as $\mathbf{M}_{3,3}^+/\alpha \cong \mathbf{M}_{3,3}$ and $\mathbf{M}_{3,3}^+ \in \mathbb{Q}(\mathbf{M}_{3,3})$. The other atom, $\beta = \mathrm{Cg}(x, y)$ is a reflection congruence as $\mathbf{M}_{3,3}^+/\beta \cong \mathbf{2}$. Note that the quasi-equation (ξ_5) is $\varepsilon_{\mathbf{M}_{3,3}^+,\beta}$. The equivalence of (4) and (5) then follows from Corollary 2.33.

It remains to show that say (5) implies (2), or equivalently, that if a modular lattice $\mathbf{L}$ fails (ξ_2), then it fails (ξ_5). Assume we are given a quadruple (x, y, u, v) of elements from L such that

$$\neg(\xi_2) \quad x \wedge (y \vee (u \wedge v)) \wedge (u \vee v) \not\leq y \vee (x \wedge u) \vee (x \wedge v).$$

We will make a series of substitutions, until we obtain a new failure (x', y', u', v') of (ξ_2) such that the new elements generate a sublattice of $\mathbf{L}$ isomorphic to either $\mathbf{M}_{3,3}$ or $\mathbf{M}_{3,3}^+$, in which case they fail (ξ_5). To simplify the calculations and notation, we make the substitutions one at a time, and avoid subscripts and superscripts on the variables.

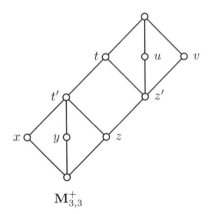

$$\mathbf{M}_{3,3}^{+}$$

FIGURE 2.6. The lattice $\mathbf{M}_{3,3}^{+}$ with labels used in the proof of Theorem 2.40.

Step 1. It is given that (x, y, u, v) is a failure of (ξ_2). Substitute $x \mapsto x' = x \wedge (u \vee v)$. Check that (x', y, u, v) is a failure of (ξ_2), because the substitution fixes the left-hand side of (ξ_2) and lowers the right-hand side. In addition, the new quadruple satisfies (dropping the prime)

$$(P_1) \quad x \leq u \vee v.$$

Step 2. Let (x, y, u, v) be a failure of (ξ_2) satisfying (P_1). Substitute $y \mapsto y' = y \wedge (u \vee v)$. Then (x, y', u, v) is a failure of (ξ_2) that still satisfies (P_1), and in addition (dropping the prime)

$$(P_2) \quad y \leq u \vee v.$$

Step 3. Let (x, y, u, v) be a failure of (ξ_2) satisfying (P_1)–(P_2). Substitute $u \mapsto u' = u \vee (x \wedge y)$. Then (x, y, u', v) fails (ξ_2), still satisfies (P_1)–(P_2), and in addition (dropping the prime)

$$(P_3) \quad u \geq x \wedge y.$$

Step 4. Let (x, y, u, v) be a failure of (ξ_2) satisfying (P_1)–(P_3). Substitute $v \mapsto v' = v \vee (x \wedge y)$. Then (x, y, u, v') fails (ξ_2), still satisfies (P_1)–(P_3), and in addition (dropping the prime)

$$(P_4) \quad v \geq x \wedge y.$$

Step 5. Let (x, y, u, v) be a failure of (ξ_2) satisfying (P_1)–(P_4). Substitute $x \mapsto x' = x \wedge (y \vee (u \wedge v))$. Then (x', y, u, v) fails (ξ_2), still satisfies (P_1)–(P_4), and in addition (dropping the prime)

$$(P_5) \quad x \leq y \vee (u \wedge v).$$

Step 6. Let (x, y, u, v) be a failure of (ξ_2) satisfying (P_1)–(P_5). Substitute $y \mapsto y' = y \wedge (x \vee (u \wedge v))$. Then (x, y', u, v) fails (ξ_2), still satisfies (P_1)–(P_5), and in addition (dropping the prime)

$$(P_6) \quad y \leq x \vee (u \wedge v).$$

Step 7. Let (x, y, u, v) be a failure of (ξ_2) satisfying (P_1)–(P_6). Substitute $u \mapsto u' = u \vee (v \wedge (x \vee y))$. Then (x, y, u', v) fails (ξ_2), still satisfies (P_1)–(P_6), and in addition (dropping the prime)

$$(P_7) \quad u \geq v \wedge (x \vee y).$$

Verifying that (ξ_2) still fails after Step 7 requires some care. Clearly the substitution $v \mapsto v'$ can only raise the left-hand side of (ξ_2). To see that the right-hand side is unchanged, use (P_6) and modularity:

$$
\begin{aligned}
y \vee (x \wedge u') \vee (x \wedge v) &= y \vee (x \wedge (u \vee (v \wedge (x \vee y)))) \vee (x \wedge v) \\
&\leq y \vee (x \wedge (u \vee (v \wedge (x \vee (u \wedge v)))))) \vee (x \wedge v) \\
&= y \vee (x \wedge (u \vee (v \wedge x) \vee (u \wedge v))) \vee (x \wedge v) \\
&= y \vee (x \wedge (u \vee (v \wedge x))) \vee (x \wedge v) \\
&= y \vee (x \wedge u) \vee (v \wedge x) \vee (x \wedge v) \\
&= y \vee (x \wedge u) \vee (x \wedge v).
\end{aligned}
$$

Step 8. Let (x, y, u, v) be a failure of (ξ_2) satisfying (P_1)–(P_7). Substitute $v \mapsto v' = v \vee (u \wedge (x \vee y))$. Then (x, y, u, v') fails (ξ_2), still satisfies (P_1)–(P_7), and in addition (dropping the prime)

$$(P_8) \quad v \geq u \wedge (x \vee y).$$

The calculation is similar to Step 7.

Step 9. Let (x, y, u, v) be a failure of (ξ_2) satisfying (P_1)–(P_8). Substitute $x \mapsto x' = x \vee (y \wedge u \wedge v)$. Then (x', y, u, v) fails (ξ_2), still satisfies (P_1)–(P_8), and in addition (dropping the prime)

$$(P_9) \quad x \geq y \wedge u \wedge v.$$

Step 10. Let (x, y, u, v) be a failure of (ξ_2) satisfying (P_1)–(P_9). Substitute $y \mapsto y' = y \vee (x \wedge u \wedge v)$. Then (x, y', u, v) fails (ξ_2), still satisfies (P_1)–(P_9), and in addition (dropping the prime)

$$(P_{10}) \quad y \geq x \wedge u \wedge v.$$

Step 11. Let (x, y, u, v) be a failure of (ξ_2) satisfying (P_1)–(P_{10}). Substitute $u \mapsto u' = u \wedge (x \vee y \vee v)$. Then (x, y, u', v) fails (ξ_2), still satisfies (P_1)–(P_{10}), and in addition (dropping the prime)

$$(P_{11}) \quad u \leq x \vee y \vee v.$$

Step 12. Let (x, y, u, v) be a failure of (ξ_2) satisfying (P_1)–(P_{11}). Substitute $v \mapsto v' = v \wedge (x \vee y \vee u)$. Then (x, y, u, v') fails (ξ_2), still satisfies (P_1)–(P_{11}), and in addition (dropping the prime)

$$(P_{12}) \quad v \leq x \vee y \vee u.$$

Now, with (x, y, u, v) a failure of ξ_2 such that (P_1)–(P_{12}) hold, let

$$z = u \wedge v \wedge (x \vee y)$$
$$t = (u \wedge v) \vee x \vee y$$

$$z' = u \wedge v$$
$$t' = x \vee y.$$

We claim that x, y, u, v generate a sublattice of $\mathbf{L}$ isomorphic to $\mathbf{M}_{3,3}^+$, as in Figure 2.6, or $\mathbf{M}_{3,3}$ if $z = z'$ and $t = t'$. Indeed, the conditions (P_1)–(P_{12}) ensure that $\{x, y, z, x \wedge y, x \vee y\}$ and $\{u, v, z, u \wedge v, u \vee z\}$ are sublattices isomorphic to $\mathbf{M}_3$, and the definition of z, t, z', t' takes care of the rest. Since (ξ_2) fails we have $x \neq y$. Thus $(x, y, z, z', u, v, t, t')$ witnesses a failure of (ξ_5). This completes the proof. $\qquad\square$

On the other hand, Grätzer and Lakser showed that the interval $[\mathbb{Q}(\mathbf{M}_{3,3}^+), \mathbb{Q}(\mathbf{M}_{3,3})]$ in $\mathrm{L_q}(\mathfrak{M})$ is uncountable. Indeed, a refinement of their proof by Adams and Dziobiak [1, 54] shows that $\mathbb{Q}(\mathbf{M}_{3,3})$ is Q-universal; see Section 5.5.

THEOREM 2.41. *The quasivariety* $\mathbb{Q}(\mathbf{M}_{3,3})$ *is Q-universal. The interval* $[\mathbb{Q}(\mathbf{M}_{3,3}^+), \mathbb{Q}(\mathbf{M}_{3,3})]$ *in* $\mathrm{L_q}(\mathfrak{M})$ *contains* $2^{\aleph_0}$ *subquasivarieties.*

2.5. Quasivarieties of Abelian Groups

The variety $\mathcal{A}$ of abelian groups provides another simple example of the results in this chapter. The variety is not locally finite, but it has finite type, and results such as Corollary 2.6 apply to its finite members. Quasivarieties of abelian groups were first described by Vinogradov [166].

Every quasivariety is determined by its finitely generated members, and every finitely generated abelian group is a direct sum of copies of the integers $\mathbf{Z}$ and prime-power cyclic groups $\mathbf{Z}_{p^k}$. The completely join irreducible subquasivarieties in $\mathrm{L_q}(\mathcal{A})$ are either $\langle \mathbf{Z}_{p^k} \rangle$ for some prime power or $\langle \mathbf{Z} \rangle$. The latter is join irreducible because every nontrivial subgroup of $\mathbf{Z}$ is isomorphic to itself. Each quasivariety of abelian groups is determined by which of the groups $\mathbf{Z}_{p^k}$ and $\mathbf{Z}$ it contains.

Note that $\mathbf{Z}_{p^k} \leq \mathbf{Z}_{p^{k+1}}$, so for each prime p the quasivarieties of this form are an ascending chain in $\mathrm{L_q}(\mathcal{A})$. Each quasivariety $\langle \mathbf{Z}_{p^k} \rangle$ is (completely) join prime, and its semi-splitting quasi-equation is

$$p^k x \approx 0 \to p^{k-1} x \approx 0.$$

Of course the quasivariety $\langle p^k x \approx 0 \to p^{k-1} x \approx 0 \rangle$ consists of all abelian groups with no element of order p^k, including all torsion-free abelian groups.

On the other hand, $\langle \mathbf{Z} \rangle$ is finitely join prime, but not completely join prime. Indeed, $\langle \mathbf{Z} \rangle \leq \bigvee_{q \in Q} \langle \mathbf{Z}_q \rangle$ for any infinite collection Q of prime powers.

Let us say that a set S of integers is *hereditary* if $m \mid n$ and $n \in S$ implies $m \in S$.

THEOREM 2.42. *Every quasivariety of abelian groups is determined by the prime-power or infinite cyclic groups it contains. For a quasivariety* $\mathcal{Q} \leq \mathcal{A}$, *this is either*

(1) $\{\mathbf{Z}_q : q \in S\}$ for a finite hereditary set S of prime powers, or

(2) $\{\mathbf{Z}\} \cup \{\mathbf{Z}_q : q \in T\}$ for an arbitrary hereditary set T of prime powers.

An easy consequence of this theorem is that $L_q(\mathcal{A})$ is a distributive lattice. Looking ahead to Section 5.5, it follows that $\mathcal{A}$ is not Q-universal, since $L_q(\mathcal{K})$ satisfies no nontrivial lattice identity when $\mathcal{K}$ is a Q-universal quasivariety, by Lemma 5.30 (Adams and Dziobiak [1]).

The results in Kearnes [103] may be regarded as a far-reaching generalization of the analysis in this section.

2.6. Quasivarieties of Infinite Type

The structure of the lattice of subquasivarieties is not nearly as nice when the type is not finite. Not surprisingly, in view of the proof of Lemma 2.4, when the type of $\mathcal{K}$ is infinite, a finitely generated quasivariety $\mathbb{Q}(\mathbf{T})$ need not be compact in $L_q(\mathcal{K})$. When a variety of algebras $\mathcal{W}$ is not locally finite, a finitely generated variety $\mathbb{V}(\mathbf{T})$ need not be compact in $L_v(\mathcal{W})$.

In order not to confuse ourselves any more than necessary, we organize this section rather carefully. It is convenient to consider general structures, with a type that may have operations and/or relations. We begin with definitions.

A structure $\mathbf{T}$ is Q-compact in a quasivariety $\mathcal{K}$ if the subquasivariety $\mathbb{Q}(\mathbf{T}) = \langle \mathbf{T} \rangle$ is compact in the lattice $L_q(\mathcal{K})$. Equivalently, $\mathbf{T}$ is Q-compact in $\mathcal{K}$ if whenever $\mathbf{T} \in \mathbb{Q}(\{\mathbf{B}_i : i \in I\})$ for a set of structures $\mathbf{B}_i \in \mathcal{K}$, then there exists a finite subset $F \subseteq I$ such that $\mathbf{T} \in \mathbb{Q}(\{\mathbf{B}_i : i \in F\})$.

Similarly, a structure $\mathbf{T}$ is V-compact in a variety $\mathcal{W}$ if the subvariety $\mathbb{V}(\mathbf{T})$ is compact in $L_v(\mathcal{W})$. Thus $\mathbf{T}$ is V-compact in $\mathcal{W}$ if whenever $\mathbf{T} \in \mathbb{V}(\{\mathbf{B}_i : i \in I\})$ for a set of structures $\mathbf{B}_i \in \mathcal{W}$, then $\mathbf{T} \in \mathbb{V}(\{\mathbf{B}_i : i \in F\})$ for some finite subset $F \subseteq I$.

The question we want to address is: *For $* \in \{Q, V\}$, under what conditions on $\mathcal{K}$ or $\mathcal{W}$ does $\mathbf{T}$ finite imply that $\mathbf{T}$ is $*$-compact?* The results are summarized in the table of Table 2.1.

Observation: If $\mathbf{T}$ is V-compact in $\mathcal{W}$, then $\mathbf{T}$ is Q-compact in any quasivariety $\mathcal{K} \subseteq \mathcal{W}$. This is just because $\mathbb{Q}(\mathcal{X}) \subseteq \mathbb{V}(\mathcal{X})$.

In this terminology, Theorem 2.4 takes the following form.

THEOREM 2.43. *If $\mathbf{T}$ is a finite structure in a quasivariety $\mathcal{K}$ of finite type, then $\mathbf{T}$ is Q-compact in $\mathcal{K}$.*

The analogous result for varieties concerns local finiteness.

THEOREM 2.44. *Let $\mathcal{W}$ be a locally finite variety, and assume that the type of $\mathcal{W}$ has only finitely many relations (and arbitrarily many operations). Then every finite structure $\mathbf{T} \in \mathcal{W}$ is V-compact in $\mathcal{W}$.*

PROOF. First consider the case when $\mathcal{W}$ is a variety of algebras, without relations (other than $\approx$) in the type. Assume $\mathbf{T} \in \mathbb{V}(\{\mathbf{B}_i : i \in I\})$ with each $\mathbf{B}_i \in \mathcal{W}$. Then there exist a sequence of structures $(\mathbf{C}_j)_{j \in J}$ such that $\{\mathbf{C}_j : j \in J\} \subseteq \{\mathbf{B}_i : i \in I\}$, a subalgebra $\mathbf{S}$ of $\prod_{j \in J} \mathbf{C}_j$ with an embedding

$f : \mathbf{S} \leq \prod_{j \in J} \mathbf{C}_j$, and a surjective homomorphism $h : \mathbf{S} \twoheadrightarrow \mathbf{T}$. Let $\{s_1, \ldots, s_n\}$ be a complete set of preimages of the elements of $\mathbf{T}$, where $n = |T|$. Since $\mathcal{W}$ is locally finite, $\mathbf{S}' = \mathrm{Sg}(\{s_1, \ldots, s_n\})$ is a finite subalgebra of $\prod_{j \in J} \mathbf{C}_j$. Since $\mathbf{S}'$ is finite, there is a finite set of indices $F \subseteq J$ such that $\pi_F f : \mathbf{S}' \leq \prod_{j \in F} \mathbf{C}_j$ is one-to-one, and hence an embedding. Moreover, the restriction $h|_{\mathbf{S}'} : \mathbf{S}' \to \mathbf{T}$ is surjective, so $\mathbf{T}$ is in the variety generated by $\{\mathbf{C}_j : j \in F\}$, which is a finite subset of the $\mathbf{B}_i$'s. Thus $\mathbf{T}$ is V-compact in $\mathcal{W}$.

Now suppose that the type of $\mathcal{W}$ contains finitely many relations. Recall that for general structures, a homomorphism $g : \mathbf{A} \to \mathbf{B}$ satisfies $g(R^{\mathbf{A}}) \subseteq R^{\mathbf{B}}$ for each relation R in the type. In our situation, this means that if R is a k-ary relation symbol and $(t_1, \ldots, t_k) \notin R^{\mathbf{T}}$, then for any $s_1, \ldots, s_k \in S$ with $h(s_i) = t_i$ for $i \leq k$, we have $(s_1, \ldots, s_k) \notin R^{\mathbf{S}}$. That, in turn, means there exists $j \in J$ such that $(s_{1j}, \ldots, s_{kj}) \notin R^{\mathbf{C}_j}$. Thus we need to enlarge F to a set $G \supseteq F$ such that whenever $s_1, \ldots, s_k \in S'$ and $(s_1, \ldots, s_k) \notin R^{\mathbf{S}'}$, there is an index $j \in G$ witnessing this fact. Since $\mathbf{S}'$ is finite and the type has only finitely many relation symbols, this can be done keeping G finite. With this modification, the proof goes through exactly as for algebras. $\square$

COROLLARY 2.45. *If $\mathcal{W}$ is a locally finite variety of finite type, then the lattice of subvarieties $\mathrm{L_v}(\mathcal{W})$ is both algebraic and dually algebraic.*

The proof mimics that of Theorem 2.5.

The analogue for varieties of a *quasicritical* structure is the notion of a *critical* structure. A structure $\mathbf{T}$ is *critical* if it is finite and not in the variety generated by the proper homomorphic images of substructures of $\mathbf{T}$, in symbols, $\mathbf{T} \notin (\mathbb{HS} \setminus \mathbb{I})(\mathbf{T})$. The role of critical structures in locally finite varieties is summarized in Neumann [134], Sections 5.1 and 5.4.

- A locally finite variety is generated by its critical members.
- Every completely join irreducible subvariety of a locally finite variety is generated by a critical structure.
- The variety generated by a critical structure need not be join irreducible.
- Distinct critical structures can generate the same variety.
- Every finite structure is in the variety generated by its critical subfactors.

Critical groups played an important part in the proof of the finite basis theorem for finite groups, Oates and Powell [140].

Let us look at some examples to show that the conditions in Theorems 2.43 and 2.44 are needed.

Example 1: *varieties $\mathcal{W}$ of finite type but not locally finite, with a finite algebra $\mathbf{T}$ not V-compact.* In the variety $\mathcal{A}$ of abelian groups, for a prime p, $\mathbf{Z}_p$ is in the variety generated by $\{\mathbf{Z}_q : q$ prime, $q \neq p\}$. In the variety $\mathcal{L}$

of lattices, the diamond $\mathbf{M}_3$ is in the variety generated by all finite (lower) bounded lattices, but not in the variety generated by any finite set of them [48].

Example 2: *a quasivariety* $\mathfrak{Q}$ *locally finite but having infinitely many relations in the type, with a finite structure* $\mathbf{T}$ *not Q-compact.* Pure relational structures, which we consider in more detail in Chapter 7, provide an easy example. Let the type of $\mathfrak{Q}$ consist of countably many unary predicates P_i ($i \in \omega$), and let $\mathfrak{Q}$ satisfy $x \approx y$, so that $\mathfrak{Q}$ contains only 1-element structures (with different sets of predicates holding); cf. Theorem A.5. In $\mathfrak{Q}$, for $k \in \omega$ let $\mathbf{B}_k$ be the structure such that $P_i x$ holds in $\mathbf{B}_k$ iff $i \geq k$, and let $\mathbf{B}_\omega$ be the structure such that every P_i ($i \in \omega$) is empty. Then $\mathbf{B}_\omega \cong \prod_{k \in \omega} \mathbf{B}_k$, so $\mathbf{B}_\omega \in \bigvee_{k \in \omega} \langle \mathbf{B}_k \rangle$, but no finite subset suffices. Indeed, the structures in any finite collection $\{\mathbf{B}_{k_1}, \ldots, \mathbf{B}_{k_m}\}$ all satisfy the atomic formula $P_n x$ for every $n \geq \max(k_1, \ldots, k_m)$, which $\mathbf{B}_\omega$ does not. Hence $\mathbf{B}_\omega \notin \bigvee_{i \leq m} \langle \mathbf{B}_{k_i} \rangle$. Thus $\langle \mathbf{B}_\omega \rangle$ is not compact in $\mathrm{L_q}(\mathfrak{Q})$.

Example 3: *a quasivariety* $\mathcal{R}$ *of algebras, not locally finite, having no relations but infinitely many operations, with a finite algebra* $\mathbf{T}$ *not Q-compact.* The idea of Example 2 can be converted to an example using 2-element algebras. Let the type of $\mathcal{R}$ consist of unary functions f_i ($i \in \omega$). For $k \in \omega$, let $\mathbf{C}_k$ be the algebra with base set $\{a, b\}$ and operations

$$f_i^{\mathbf{C}_k}(a) = b \qquad \text{for all } i,$$

$$f_i^{\mathbf{C}_k}(b) = \begin{cases} a & \text{if } i < k, \\ b & \text{if } i \geq k. \end{cases}$$

In $\prod_{k \in \omega} \mathbf{C}_k$, let $\mathbf{S}$ be the subalgebra consisting of all sequences that are eventually constant. It is easy to see that $\mathbf{S}$ is a subalgebra of $\prod_{k \in \omega} \mathbf{C}_k$. Let F be the filter on ω consisting of all subsets $\uparrow n = \{k : k \geq n\}$, so that the congruence $\equiv_F$ on $\prod_{k \in \omega} \mathbf{C}_k$ identifies all sequences that are eventually equal. Restricted to $\mathbf{S}$ we have two congruence classes, $[a]$ containing all sequences that are eventually a, and $[b]$ containing all sequences that are eventually b. The algebra $\mathbf{C}_\omega := \mathbf{S}/\equiv_F$ has

$$f_i^{\mathbf{C}_\omega}[a] = [b] \qquad f_i^{\mathbf{C}_\omega}[b] = [a]$$

for all i. Again $\mathbf{C}_\omega \in \bigvee_{k \in \omega} \langle \mathbf{C}_k \rangle$, since it is a subalgebra of a reduced product, but no finite subset suffices. For the algebras in any finite collection $\{\mathbf{C}_{k_1}, \ldots, \mathbf{C}_{k_m}\}$ all satisfy $f_n x \approx f_n y$ for $n \geq \max(k_1, \ldots, k_m)$, while $\mathbf{C}_\omega$ fails that equation. Therefore $\mathbf{C}_\omega \notin \bigvee_{i \leq m} \langle \mathbf{C}_{k_i} \rangle$, and we conclude that $\langle \mathbf{C}_\omega \rangle$ is not compact in $\mathrm{L_q}(\mathcal{R})$.

The theorems and examples in this section, with the observation that V-compact implies Q-compact, are summarized in Table 2.1.

	finite type	loc. fin., fin. m. rel.	loc. fin., inf. m. rel.	not loc. fin., fin. m. rel.
Q-compact	Yes, Th 2.43	Yes, Obsv	No, Ex 2	No, Ex 3
V-compact	No, Ex 1	Yes, Th 2.44	No, Obsv	No, Obsv

TABLE 2.1. Q-compactness *versus* V-compactness. The table summarizes the conditions on a variety or quasivariety $\mathcal{K}$ under which every finite structure $\mathbf{T}$ in $\mathcal{K}$ is Q-compact or V-compact. The abbreviations are as follows. "loc. fin." is *locally finite*, "fin. m. rel." is *finitely many relations*, "inf. m. rel." is *infinitely many relations*. "Th" refers to the theorem number, "Ex" to the example number, and "Obsv" to the observation above.

CHAPTER 3

Omission and Bases for Quasivarieties

3.1. Characterizing Quasivarieties by Excluded Subalgebras

In this section we show that if $\mathcal{K}$ is a locally finite quasivariety of finite type, $\mathbf{T}$ is a finite algebra in $\mathcal{K}$, and $\alpha \succ \Delta$ in $\mathrm{Con}_{\mathcal{K}} \mathbf{T}$, then the quasivariety $\langle \varepsilon_{\mathbf{T},\alpha} \rangle$ consists of all algebras in $\mathcal{K}$ that omit a finite list of forbidden subalgebras. A slight variation finds the quasivarieties that are minimal with respect to not being contained in $\langle \varepsilon_{\mathbf{T},\alpha} \rangle$. Both these results are in Theorem 3.4. As a consequence, subquasivarieties that are finitely based relative to $\mathcal{K}$ can be characterized by the exclusion of finitely many subalgebras (Theorem 3.10).

The first observation just restates what it means to satisfy $\varepsilon_{\mathbf{T},\alpha}$.

LEMMA 3.1. *An algebra $\mathbf{A}$ satisfies $\varepsilon_{\mathbf{T},\alpha}$ if and only if $\ker h \geq \alpha$ for every homomorphism $h : \mathbf{T} \to \mathbf{A}$.*

This gives us a useful method for testing whether $\mathbf{A}$ satisfies $\varepsilon_{\mathbf{T},\alpha}$. From the congruence lattice of $\mathbf{T}$, one can find all the homomorphic images of $\mathbf{T}$ with $h(\mathbf{T}) \in \mathcal{K}$ and $\ker h \not\geq \alpha$. Form a list of these: $\mathbf{Z}_1, \ldots, \mathbf{Z}_n$. If $\mathbf{Z}_j \leq \mathbf{A}$ for some j, then $\mathbf{A}$ fails $\varepsilon_{\mathbf{T},\alpha}$. Otherwise, $\mathbf{A}$ satisfies $\varepsilon_{\mathbf{T},\alpha}$. (The algebras $\mathbf{Z}_1, \ldots, \mathbf{Z}_n$ comprise the set $\mathbf{B}(\mathbf{T}, \alpha)$ defined below.)

These observations in turn lead to results about excluded subalgebras. Let us record first the crude form, which is interesting in itself, before going to refined versions.

THEOREM 3.2. *Let $\mathcal{K}$ be a locally finite quasivariety of finite type. Let*

$$\varepsilon : \quad \&_i \, s_i \approx t_i \to u \approx v$$

be a quasi-equation in the language of $\mathcal{K}$. Then there exist finite algebras $\mathbf{B}_1, \ldots, \mathbf{B}_k \in \mathcal{K}$ such that for all $\mathbf{A} \in \mathcal{K}$, $\mathbf{A}$ satisfies ε if and only if $\mathbf{B}_j \not\leq \mathbf{A}$ for all j.

PROOF. Let $\mathbf{F}$ be the $\mathcal{K}$-free algebra generated by the variables of ε. By hypothesis, $\mathbf{F}$ is finite. In $\mathrm{Con}_{\mathcal{K}} \mathbf{F}$, let $\eta = \bigvee_i \mathrm{Cg}_{\mathcal{K}}(s_i, t_i)$, and let $\tau = \mathrm{Cg}_{\mathcal{K}}(u, v)$. Let

$$\mathcal{B} = \{ \theta \in \mathrm{Con}_{\mathcal{K}} \mathbf{F} : \theta \geq \eta, \; \theta \not\geq \tau \}$$

and let $\{ \mathbf{B}_1, \ldots, \mathbf{B}_k \} = \{ \mathbf{F}/\theta : \theta \in \mathcal{B} \}$. (We may if desired take only those $\mathbf{B}_j$ that are minimal with respect to the embedding order $\leq$.)

J. Hyndman, J. B. Nation, *The Lattice of Subquasivarieties of a Locally Finite Quasivariety*, CMS Books in Mathematics, https://doi.org/10.1007/978-3-319-78235-5_3

An algebra $\mathbf{A} \in \mathcal{K}$ satisfies ε if and only if for every homomorphism $h : \mathbf{F} \to \mathbf{A}$, we have $\ker h \geq \eta$ implies $\ker h \geq \tau$. Hence $\mathbf{A}$ satisfies ε if and only if $\mathbf{B}_j \not\leq \mathbf{A}$ for all j.   □

Let us refine this observation. There are two natural orders connected with algebras in $\mathcal{K}$:

- $\mathbf{A} \leq \mathbf{B}$ if $\mathbf{A}$ is a subalgebra of $\mathbf{B}$,
- $\mathbf{A} \leq_q \mathbf{B}$ if $\langle \mathbf{A} \rangle \leq \langle \mathbf{B} \rangle$, i.e., $\mathbf{A} \leq \mathbf{B}^n$ for some n.

The latter is in general only a quasi-order, but a partial order when restricted to finite quasicritical algebras in a locally finite quasivariety $\mathcal{K}$ of finite type, by Theorem 2.11. Of course, $\mathbf{A} \leq \mathbf{B}$ implies $\mathbf{A} \leq_q \mathbf{B}$.

With each quasivariety $\langle \varepsilon_{\mathbf{T},\alpha} \rangle$ we associate three finite sets of finite algebras.

- $\mathbf{B}(\mathbf{T}, \alpha) = \{\mathbf{Z}_1, \dots, \mathbf{Z}_n\}$ consists of all $\mathbf{T}/\theta$ with $\theta \not\geq \alpha$ in $\mathrm{Con}_{\mathcal{K}} \mathbf{T}$.
- $\mathbf{O}(\mathbf{T}, \alpha) = \{\mathbf{U}_1, \dots, \mathbf{U}_k\}$ consists of the algebras that are minimal in $(\mathbf{B}(\mathbf{T}, \alpha), \leq)$.
- $\mathbf{N}(\mathbf{T}, \alpha) = \{\mathbf{R}_1, \dots, \mathbf{R}_m\}$ consists of the quasicritical algebras that are minimal in $(\mathbf{B}(\mathbf{T}, \alpha), \leq_q)$.

Note that it is relatively straightforward to find these sets of algebras from the labeled lattice of $\mathcal{K}$-congruences of $\mathbf{T}$. It may even happen that $\mathbf{O}(\mathbf{T}, \alpha)$ contains only $\mathbf{T}$ itself, as happens with $\mathcal{K}$-subdirectly irreducible algebras (see Lemma 3.6). For complicated cases, the Universal Algebra Calculator can be of assistance. The UA Calculator was used to compute some of the congruence lattices in Chapter 5; see Figures 5.4, 5.8, 5.10, 5.11, 5.12, 5.13, 5.16, and 5.19.

It is possible that $\langle \mathbf{Z} \rangle = \langle \mathbf{Z}' \rangle$ for distinct $\mathbf{Z}, \mathbf{Z}' \in \mathbf{B}(\mathbf{T}, \alpha)$, not both quasicritical. Likewise, it may be that $\langle \mathbf{U} \rangle \leq \langle \mathbf{U}' \rangle$ for distinct $\mathbf{U}, \mathbf{U}' \in \mathbf{O}(\mathbf{T}, \alpha)$, e.g., if $\mathbf{U} \leq \mathbf{U}' \times \mathbf{U}'$. The next lemma addresses these technical difficulties.

LEMMA 3.3. *For any finite algebra* $\mathbf{T} \in \mathcal{K}$ *and* $\alpha \succ \Delta$ *in* $\mathrm{Con}_{\mathcal{K}} \mathbf{T}$, *the following hold.*

(1) *For any* $\mathbf{Z} \in \mathbf{B}(\mathbf{T}, \alpha)$, *there exists a quasicritical algebra* $\mathbf{Z}' \in \mathbf{B}(\mathbf{T}, \alpha)$ *such that* $\mathbf{Z}' \leq \mathbf{Z}$.

(2) *If* $\mathbf{U} \in \mathbf{O}(\mathbf{T}, \alpha)$, *then* $\mathbf{U}$ *is quasicritical.*

Hence $\mathbf{N}(\mathbf{T}, \alpha) \subseteq \mathbf{O}(\mathbf{T}, \alpha) \subseteq \mathbf{B}(\mathbf{T}, \alpha)$.

PROOF. If $\mathbf{Z} \in \mathbf{B}(\mathbf{T}, \alpha)$ is not quasicritical, then it is a subdirect product of its proper subalgebras, and at least one of those factors $\mathbf{Z}'$ is not above α in $\mathrm{Con}_{\mathcal{K}} \mathbf{T}$. If $\mathbf{U} \in \mathbf{O}(\mathbf{T}, \alpha)$ were not quasicritical, then applying the same argument to $\mathbf{U}$ would contradict the minimality of $\mathbf{U}$ in the subalgebra order.   □

With these sets in hand, Theorem 3.2 takes the following useful form.

THEOREM 3.4. *Let* $\mathcal{K}$ *be a locally finite quasivariety of finite type. Let* $\mathbf{T}$ *be a finite algebra in* $\mathcal{K}$ *and* $\alpha \succ \Delta$ *in* $\mathrm{Con}_{\mathcal{K}} \mathbf{T}$.

(1) *For any algebra* $\mathbf{A} \in \mathcal{K}$, *we have* $\mathbf{A} \in \langle \varepsilon_{\mathbf{T},\alpha} \rangle$ *if and only if* $\mathbf{A}$ *omits* $\mathbf{U}$ *as a subalgebra for all* $\mathbf{U} \in \mathbf{O}(\mathbf{T}, \alpha)$.

(2) *For any quasivariety* $\mathcal{Q} \leq \mathcal{K}$, *we have* $\mathcal{Q} \leq \langle \varepsilon_{\mathbf{T},\alpha} \rangle$ *if and only if* $\mathbf{R} \notin \mathcal{Q}$ *for all* $\mathbf{R} \in \mathbf{N}(\mathbf{T}, \alpha)$.

Let us consider several consequences. We can begin by adding to the equivalences of Corollary 2.33.

COROLLARY 3.5. *Let* $\mathbf{T}$ *be a finite algebra in* $\mathcal{K}$. *Then* $\langle \mathbf{T} \rangle$ *is completely join prime in* $\mathrm{L}_q(\mathcal{K})$ *if and only if there is an atom* $\alpha \in \mathrm{Con}_{\mathcal{K}} \mathbf{T}$ *such that* $\mathbf{N}(\mathbf{T}, \alpha) = \{\mathbf{T}\}$.

Looking ahead to the quasivariety $\mathcal{M}$ of Chapter 5 we can see this phenomenon in Table 5.4, where the quasivarieties generated by the algebras $\mathbf{T}_1$, $\mathbf{T}_3$, $\mathbf{T}_4$, $\mathcal{F}_1$, $\mathbf{R}_6$, $\mathbf{R}_8$ and $\mathbf{P}_2$ have $\mathbf{N}(\mathbf{T}, \alpha) = \{\mathbf{T}\}$. An even stronger condition is that $\mathbf{O}(\mathbf{T}, \alpha) = \{\mathbf{T}\}$, enjoyed by $\mathbf{T}_1$, $\mathbf{T}_3$, $\mathbf{T}_4$, and $\mathbf{R}_6$ in that table.

We say that an algebra $\mathbf{T} \in \mathcal{K}$ is $\mathcal{K}$-*subdirectly irreducible* if $\mathrm{Con}_{\mathcal{K}} \mathbf{T}$ has a unique atom, called the $\mathcal{K}$-*monolith*. Finite $\mathcal{K}$-subdirectly irreducible algebras are the analogue for quasivarieties of finite projective subdirectly irreducible algebras for varieties, in the sense that if $\mathbf{T}$ is finite and $\mathcal{K}$-subdirectly irreducible, then $\mathbf{T} \in \mathbb{SPU}(\mathcal{X})$ for a collection of algebras $\mathcal{X} \subseteq \mathcal{K}$, then $\mathbf{T} \leq \mathbf{B}$ for some $\mathbf{B} \in \mathcal{X}$. (The argument resembles the proof of Lemma 2.4.)

COROLLARY 3.6. *The following are equivalent for a finite algebra in a locally finite quasivariety* $\mathcal{K}$ *of finite type.*

(1) $\mathbf{T}$ *is* $\mathcal{K}$-*subdirectly irreducible.*

(2) $\mathbf{O}(\mathbf{T}, \alpha) = \{\mathbf{T}\}$ *for some atom* $\alpha \in \mathrm{Con}_{\mathcal{K}} \mathbf{T}$.

Note that this implies $\mathbf{N}(\mathbf{T}, \alpha) = \{\mathbf{T}\}$. *Moreover, if* $\mathbf{T}$ *is* $\mathcal{K}$-*subdirectly irreducible with* $\mathcal{K}$-*monolith* α, *then for any algebra* $\mathbf{A} \in \mathcal{K}$ *we have that* $\mathbf{A} \in \langle \varepsilon_{\mathbf{T},\alpha} \rangle$ *if and only if* $\mathbf{A}$ *omits* $\mathbf{T}$ *as a subalgebra.*

PROOF. To see that (1) implies (2), note that the only congruence in $\mathrm{Con}_{\mathcal{K}} \mathbf{T}$ not above α is Δ, the identity congruence, and $\mathbf{T}/\Delta \cong \mathbf{T}$. Hence $\mathbf{O}(\mathbf{T}, \alpha) = \{\mathbf{T}\}$.

Conversely, assume $\mathbf{O}(\mathbf{T}, \alpha) = \{\mathbf{T}\}$, so that for algebras $\mathbf{S} \in \mathcal{K}$ we have $\mathbf{S}$ satisfies $\varepsilon_{\mathbf{T},\alpha}$ if and only if $\mathbf{T} \not\leq \mathbf{S}$. Then $\mathbf{T}/\alpha$ satisfies $\varepsilon_{\mathbf{T},\alpha}$, so it is a reflection congruence by Corollary 2.34. If $\mathbf{T}$ had a nonzero $\mathcal{K}$-congruence θ with $\theta \not\geq \alpha$, then likewise $\mathbf{T}/\theta$ satisfies $\varepsilon_{\mathbf{T},\alpha}$. But since α is an atom of $\mathrm{Con}_{\mathcal{K}} \mathbf{T}$, then $\alpha \wedge \theta = \Delta$, whence $\mathbf{T} \leq \mathbf{T}/\alpha \times \mathbf{T}/\theta$ so that $\mathbf{T}$ satisfies $\varepsilon_{\mathbf{T},\alpha}$, a contradiction. Therefore α is the unique atom of $\mathrm{Con}_{\mathcal{K}} \mathbf{T}$, and $\mathbf{T}$ is $\mathcal{K}$-subdirectly irreducible, as desired. □

The next observations are routine but useful.

COROLLARY 3.7. *Let* $\mathcal{K}$ *be a locally finite quasivariety of finite type, and let* $\mathbf{T}$ *be a finite algebra in* $\mathcal{K}$. *Let* α *be an atom of* $\mathrm{Con}_{\mathcal{K}} \mathbf{T}$, *and let* ξ *be a* $\mathcal{K}$-*congruence maximal with respect to* $\xi \not\geq \alpha$. *Then* $\mathbf{T}/\xi$ *is* $\mathcal{K}$-*subdirectly irreducible. Moreover,* $\mathbf{T}$ *is a subdirect product of such* $\mathcal{K}$-*subdirectly irreducible factors.*

COROLLARY 3.8. *Let $\mathcal{K}$ be a locally finite quasivariety of finite type. For $i \in I$, let $\mathbf{T}_i$ be finite algebras in $\mathcal{K}$ and $\alpha_i \succ \Delta$ in $\mathrm{Con}_{\mathcal{K}} \mathbf{T}_i$.*

(1) *An algebra $\mathbf{A} \in \mathcal{K}$ satisfies $\mathbf{A} \in \bigwedge_{i \in I} \langle \varepsilon_{\mathbf{T}_i, \alpha_i} \rangle$ if and only if $\mathbf{A}$ omits $\mathbf{U}$ for all $\mathbf{U} \in \bigcup_{i \in I} \mathbf{O}(\mathbf{T}_i, \alpha_i)$. If the index set I is finite, it suffices to consider only those $\mathbf{U} \in \bigcup_{i \in I} \mathbf{O}(\mathbf{T}_i, \alpha_i)$ that are minimal with respect to subalgebra inclusion.*

(2) *A quasivariety $\mathcal{Q} \leq \mathcal{K}$ satisfies $\mathcal{Q} \leq \bigwedge_{i \in I} \langle \varepsilon_{\mathbf{T}_i, \alpha_i} \rangle$ if and only if $\mathbf{R} \nleq \mathcal{Q}$ for all $\mathbf{R} \in \bigcup_{i \in I} \mathbf{N}(\mathbf{T}_i, \alpha_i)$. If the index set I is finite, it suffices to consider only those $\mathbf{R} \in \bigcup_{i \in I} \mathbf{N}(\mathbf{T}_i, \alpha_i)$ that are minimal with respect to quasivariety inclusion.*

The formulation of Corollary 3.8(1) is applied repeatedly in Chapter 5 to describe quasivarieties $\langle \mathbf{T} \rangle$ contained in our test variety $\mathcal{M}$.

Since $\mathrm{L_q}(\mathcal{K})$ is algebraic, every subquasivariety of $\mathcal{K}$ is a meet of completely meet irreducible quasivarieties. For every completely meet irreducible quasivariety $\mathcal{M}$ in $\mathrm{L_q}(\mathcal{K})$, there are a finite quasicritical algebra $\mathbf{T}$, not in $\mathcal{M}$, and an atom α of $\mathrm{Con}_{\mathcal{K}} \mathbf{T}$ such that $\mathcal{M} = \langle \varepsilon_{\mathbf{T}, \alpha} \rangle$. If we write an arbitrary subquasivariety $\mathcal{W} \leq \mathcal{K}$ as $\mathcal{W} = \bigwedge_{i \in I} \langle \varepsilon_{\mathbf{T}_i, \alpha_i} \rangle$, Corollary 3.8(1) takes the following form.

COROLLARY 3.9. *Let $\mathcal{K}$ be a locally finite quasivariety of finite type, and let $\mathcal{W} \leq \mathcal{K}$.*

(1) *There is a set $\mathcal{O}(\mathcal{W})$ of finite quasicritical algebras, not in $\mathcal{W}$, such that an algebra $\mathbf{A} \in \mathcal{K}$ is in $\mathcal{W}$ if and only if $\mathbf{U} \nleq \mathbf{A}$ for every $\mathbf{U} \in \mathcal{O}(\mathcal{W})$.*

(2) *There is a set $\mathcal{N}(\mathcal{W})$ of finite quasicritical algebras, not in $\mathcal{W}$, such that a quasivariety $\mathcal{Q} \leq \mathcal{K}$ satisfies $\mathcal{Q} \leq \mathcal{W}$ if and only if $\mathbf{R} \nleq \mathcal{Q}$ for every $\mathbf{R} \in \mathcal{N}(\mathcal{W})$.*

We can gather some of the preceding results into a characterization of finitely based quasivarieties in $\mathcal{K}$.

THEOREM 3.10. *Let $\mathcal{K}$ be a locally finite quasivariety of finite type. The following are equivalent for a quasivariety $\mathcal{Q} \subseteq \mathcal{K}$.*

(1) *$\mathcal{Q}$ is finitely based relative to $\mathcal{K}$.*

(2) *$\mathcal{Q}$ is dually compact in $\mathrm{L_q}(\mathcal{K})$.*

(3) *There are finitely many quasicritical algebras $\mathbf{T}_i \in \mathcal{K}$ and atoms α_i of $\mathrm{Con}_{\mathcal{K}} \mathbf{T}_i$ such that*

$$\mathcal{Q} = \langle \varepsilon_{\mathbf{T}_1, \alpha_1} \rangle \wedge \cdots \wedge \langle \varepsilon_{\mathbf{T}_n, \alpha_n} \rangle.$$

(4) *There are finitely many finite quasicritical algebras $\mathbf{R}_1, \ldots, \mathbf{R}_m$ such that for any quasivariety $\mathcal{W} \leq \mathcal{K}$, we have $\mathcal{W} \leq \mathcal{Q}$ if and only if $\mathbf{R}_i \nleq \mathcal{W}$ for all i.*

(5) *There are finitely many finite quasicritical algebras $\mathbf{U}_1, \ldots, \mathbf{U}_k$ such that for any algebra $\mathbf{A} \in \mathcal{K}$, we have $\mathbf{A} \in \mathcal{Q}$ if and only if $\mathbf{A}$ omits every $\mathbf{U}_j$ as a subalgebra.*

PROOF. The equivalence of (1) and (2) is due to the general duality between quasivarieties and quasi-equational theories. Moreover, since $L_q(\mathcal{K})$ is algebraic, every element is a meet of completely meet irreducible quasivarieties. These have the form $\langle \varepsilon_{\mathbf{T},\alpha} \rangle$ with $\mathbf{T}$ quasicritical, which are clearly finitely based relative to $\mathcal{K}$. The dually compact quasivarieties are just the meets of finitely many of these, adding the equivalence of (3).

Assume that (3) holds. Then for any quasivariety $\mathcal{W} \leq \mathcal{K}$, we have $\mathcal{W} \not\leq \mathcal{Q}$ if and only if $\mathcal{W} \not\leq \langle \varepsilon_{\mathbf{T}_i,\alpha_i} \rangle$ for some i. Applying Theorem 3.4 (1) yields that $\mathbf{V} \in \mathcal{W}$ for some $\mathbf{V} \in \bigcup_{j=1}^{n} \mathbf{O}(\mathbf{T}_i, \alpha_i)$, and thus (5).

Clearly, (5) implies (4). (In practice, the $\mathbf{R}_i$'s are a subset of the $\mathbf{U}_j$'s, since some of the latter could generate comparable quasivarieties.)

Now assume (4). Since $\mathbf{R}_1, \ldots, \mathbf{R}_m \notin \mathcal{Q}$, we know by Theorem 2.26 that, for each i, $\mathcal{Q}$ satisfies $\varepsilon_{\mathbf{R}_i,\alpha_i}$ for some atom α_i. (We need only consider those $\mathbf{R}_i$ such that $\langle \mathbf{R}_i \rangle$ is minimal with respect to being not below $\mathcal{Q}$, since those will be included in the list. Even then, for some i it may be that $\mathcal{Q}$ satisfies $\varepsilon_{\mathbf{R}_i,\alpha}$ and $\varepsilon_{\mathbf{R}_i,\beta}$ with $\gamma(\alpha)$ and $\gamma(\beta)$ distinct minimal reflection congruences in $\gamma(\mathbf{R}_i)$; in that case, both should be included.) Thus $\mathcal{Q} \leq \bigwedge_i \langle \varepsilon_{\mathbf{R}_i,\alpha_i} \rangle$ holds. Letting $\bigwedge_i \langle \varepsilon_{\mathbf{R}_i,\alpha_i} \rangle = \mathcal{N}$, we want to show that $\mathcal{Q} = \mathcal{N}$, so that property (3) holds.

Suppose to the contrary that $\mathcal{Q} < \mathcal{N}$. Then there is a completely meet irreducible quasivariety, which is $\langle \varepsilon_{\mathbf{S},\beta} \rangle$ for some finite quasicritical algebra $\mathbf{S}$ and atom β, such that $\langle \varepsilon_{\mathbf{S},\beta} \rangle \geq \mathcal{Q}$ but $\langle \varepsilon_{\mathbf{S},\beta} \rangle \not\geq \mathcal{N}$. Moreover, the algebras $\mathbf{V}_1, \ldots, \mathbf{V}_\ell$ of $\mathbf{N}(\mathbf{S}, \beta)$ have the property that, for any quasivariety $\mathcal{W} \leq \mathcal{K}$, we have $\mathcal{W} \not\leq \langle \varepsilon_{\mathbf{S},\beta} \rangle$ if and only if $\mathbf{V}_j \in \mathcal{W}$ for some j.

The inclusion $\mathcal{Q} \leq \langle \varepsilon_{\mathbf{S},\beta} \rangle$ implies that $\mathbf{V}_j \notin \mathcal{Q}$ for all j, whence by the assumption (4) means that for each j there exists i such that $\mathbf{R}_i \in \langle \mathbf{V}_j \rangle$. On the other hand, $\mathcal{N} \not\leq \langle \varepsilon_{\mathbf{S},\beta} \rangle$ implies that $\mathbf{V}_{j_0} \in \mathcal{N}$ for some j_0. Choose i_0 such that $\mathbf{R}_{i_0} \in \langle \mathbf{V}_{j_0} \rangle$. Combining these yields

$$\langle \mathbf{R}_{i_0} \rangle \leq \langle \mathbf{V}_{j_0} \rangle \leq \mathcal{N} \leq \langle \varepsilon_{\mathbf{R}_{i_0},\alpha_{i_0}} \rangle,$$

a contradiction. Thus $\mathcal{Q} = \bigwedge_i \langle \varepsilon_{\mathbf{R}_i,\alpha_i} \rangle$, as desired. $\square$

An anonymous referee pointed out that local finiteness plays an essential role in Theorem 3.10. Let $\mathcal{J}$ denote the quasivariety of all 1-unary algebras satisfying $fx \approx fy \rightarrow x \approx y$. This is of course finitely based, but there are infinitely many quasicritical algebras $\mathbf{D}_k$ $(k \geq 2)$ that are minimal with respect to not being in $\mathcal{J}$, consisting of a cycle C of length k and one additional point d with $d \notin C$ but $f(d) \in C$. For each of these, $\langle \mathbf{D}_k \rangle \not\leq \mathcal{J}$ but $\langle \mathbf{D}_k \rangle_* \leq \mathcal{J}$. (This last fact can be verified using Corollary 2.29.)

Next we describe the order and dual dependencies on the completely meet irreducible quasivarieties in $L_q(\mathcal{K})$. The completely meet irreducibles are of the form $\langle \varepsilon_{\mathbf{T},\alpha} \rangle$, so Theorem 3.4 applies.

Recall the notion of *dual refinement*. For subsets of a (quasi)-ordered set, we write $A \gg B$ to mean that for all $a \in A$ there exists $b \in B$ with $a \geq b$. This is not the same as $B \ll A$. When applying dual refinement to sets of algebras ordered by $\leq_q$, it is convenient to use the notation $A \gg_q B$.

THEOREM 3.11. *Let $\mathcal{K}$ be a locally finite quasivariety of finite type. Let* $\mathbf{S}, \mathbf{T} \in \mathcal{K}$ *with* $\alpha \succ \Delta$ *in* $\mathrm{Con}_{\mathcal{K}} \mathbf{T}$ *and* $\beta \succ \Delta$ *in* $\mathrm{Con}_{\mathcal{K}} \mathbf{S}$. *The following are equivalent in* $\mathrm{L_q}(\mathcal{K})$.

(1) $\langle \varepsilon_{\mathbf{T},\alpha} \rangle \le \langle \varepsilon_{\mathbf{S},\beta} \rangle$.
(2) $\mathbf{O}(\mathbf{S}, \beta) \gg \mathbf{O}(\mathbf{T}, \alpha)$, *i.e., for all* $\mathbf{V} \in \mathbf{O}(\mathbf{S}, \beta)$ *there is an algebra* $\mathbf{U} \in \mathbf{O}(\mathbf{T}, \alpha)$ *such that* $\mathbf{V} \ge \mathbf{U}$.
(3) $\mathbf{N}(\mathbf{S}, \beta) \gg_q \mathbf{N}(\mathbf{T}, \alpha)$, *i.e., for all* $\mathbf{W} \in \mathbf{N}(\mathbf{S}, \beta)$ *there is an algebra* $\mathbf{R} \in \mathbf{N}(\mathbf{T}, \alpha)$ *such that* $\langle \mathbf{W} \rangle \ge \langle \mathbf{R} \rangle$.

Of course, $\langle \mathbf{W} \rangle \ge \langle \mathbf{R} \rangle$ if and only if $\mathbf{R} \in \langle \mathbf{W} \rangle$. Rather than proving the results separately, let us extend the previous theorem to include dual dependencies $\bigwedge_{i \in I} \langle \delta_i \rangle \le \langle \varepsilon \rangle$ in $\mathrm{L_q}(\mathcal{K})$. Since completely meet irreducible quasivarieties are dually compact in $\mathrm{L_q}(\mathcal{K})$, it suffices to consider only finite index sets.

THEOREM 3.12. *Let $\mathcal{K}$ be a locally finite quasivariety of finite type. For* $i \in I$, *let* $\mathbf{S}, \mathbf{T}_i \in \mathcal{K}$ *with* $\alpha_i \succ \Delta$ *in* $\mathrm{Con}_{\mathcal{K}} \mathbf{T}_i$ *and* $\beta \succ \Delta$ *in* $\mathrm{Con}_{\mathcal{K}} \mathbf{S}$. *The following are equivalent in* $\mathrm{L_q}(\mathcal{K})$.

(1) $\bigwedge_{i \in I} \langle \varepsilon_{\mathbf{T}_i,\alpha_i} \rangle \le \langle \varepsilon_{\mathbf{S},\beta} \rangle$.
(2) $\mathbf{O}(\mathbf{S}, \beta) \gg \bigcup_{i \in I} \mathbf{O}(\mathbf{T}_i, \alpha_i)$, *i.e., for all* $\mathbf{V} \in \mathbf{O}(\mathbf{S}, \beta)$ *there exists* $\mathbf{U} \in \bigcup_{i \in I} \mathbf{O}(\mathbf{T}_i, \alpha_i)$ *such that* $\mathbf{V} \ge \mathbf{U}$.
(3) $\mathbf{N}(\mathbf{S}, \beta) \gg_q \bigcup_{i \in I} \mathbf{N}(\mathbf{T}_i, \alpha_i)$, *i.e., for all* $\mathbf{W} \in \mathbf{N}(\mathbf{S}, \beta)$ *there exists* $\mathbf{R} \in \bigcup_{i \in I} \mathbf{N}(\mathbf{T}_i, \alpha_i)$ *such that* $\langle \mathbf{W} \rangle \ge \langle \mathbf{R} \rangle$.

Since Theorem 3.12 includes Theorem 3.11, and parts (2) and (3) of each are similar, let us prove carefully only the equivalence of (1) and (2) of Theorem 3.12.

PROOF. Suppose that (1) fails, so that $\bigwedge_i \langle \varepsilon_{\mathbf{T}_i,\alpha_i} \rangle \not\le \langle \varepsilon_{\mathbf{S},\beta} \rangle$. Then there is an algebra $\mathbf{B} \in \bigwedge_i \langle \varepsilon_{\mathbf{T}_i,\alpha_i} \rangle \setminus \langle \varepsilon_{\mathbf{S},\beta} \rangle$. By Theorem 3.4, there exists $\mathbf{V} \in \mathbf{O}(\mathbf{S}, \beta)$ such that $\mathbf{V} \le \mathbf{B}$. Then $\mathbf{V}$ likewise satisfies $\mathbf{V} \in \bigwedge_i \langle \varepsilon_{\mathbf{T}_i,\alpha_i} \rangle \setminus \langle \varepsilon_{\mathbf{S},\beta} \rangle$. It follows that for all $\mathbf{U} \in \bigcup \mathbf{O}(\mathbf{T}_i, \alpha_i)$ we have $\mathbf{U} \not\le \mathbf{V}$. Thus (2) fails.

Conversely, suppose that (2) fails, so that for some $\mathbf{V} \in \mathbf{O}(\mathbf{S}, \beta)$ we have $\mathbf{U} \not\le \mathbf{V}$ whenever $\mathbf{U} \in \bigcup \mathbf{O}(\mathbf{T}_i, \alpha_i)$. Again applying Theorem 3.4, we obtain $\mathbf{V} \in \langle \varepsilon_{\mathbf{T}_i,\alpha_i} \rangle$ for all i, whence $\mathbf{V} \in \bigwedge_i \langle \varepsilon_{\mathbf{T}_i,\alpha_i} \rangle \setminus \langle \varepsilon_{\mathbf{S},\beta} \rangle$. Thus (1) fails. $\square$

The following corollary will be used in Propositions 5.25(11) and 7.9.

COROLLARY 3.13. *The meet* $\langle \varepsilon_{\mathbf{T}_1,\alpha_1} \rangle \wedge \langle \varepsilon_{\mathbf{T}_2,\alpha_2} \rangle = \langle \varepsilon_{\mathbf{S},\beta} \rangle$ *holds if and only if* $\mathbf{O}(\mathbf{S}, \beta) = \min(\mathbf{O}(\mathbf{T}_1, \alpha_1) \cup \mathbf{O}(\mathbf{T}_2, \alpha_2))$, *where* $\min(X)$ *denotes the set of minimal members of X with respect to subalgebra inclusion.*

PROOF. By the preceding theorem, $\langle \varepsilon_{\mathbf{T}_1,\alpha_1} \rangle \wedge \langle \varepsilon_{\mathbf{T}_2,\alpha_2} \rangle = \langle \varepsilon_{\mathbf{S},\beta} \rangle$ is equivalent to

$$\mathbf{O}(\mathbf{S}, \beta) \gg \mathbf{O}(\mathbf{T}_1, \alpha_1) \cup \mathbf{O}(\mathbf{T}_2, \alpha_2) \gg \mathbf{O}(\mathbf{S}, \beta).$$

Since $\mathbf{O}(\mathbf{S}, \beta)$ is an antichain, the conclusion follows. $\square$

3.2. Pseudoquasivarieties

A *pseudoquasivariety* is a collection $\mathcal{K}$ of finite structures that is closed under taking substructures and finite direct products, i.e., $\mathbb{SP}_{\text{fin}}(\mathcal{K}) = \mathcal{K}$. At least the first half of the basic theorem on pseudoquasivarieties is due to C. J. Ash [23]; see also Chapter 2 of Gorbunov [77].

THEOREM 3.14. *Let $\mathcal{K}$ be a pseudoquasivariety of structures of finite type. Fix any quasivariety $\mathcal{Y}$ with $\mathcal{K} \subseteq \mathcal{Y}$. Then the following hold.*

(1) $\mathcal{K}$ *is the set of all finite structures in the quasivariety $\mathcal{Q} = \mathbb{SPU}(\mathcal{K})$.*
(2) *There is a set $\mathbf{O}(\mathcal{K})$ of finite quasicritical algebras in $\mathcal{Y}$ such that for any finite algebra $\mathbf{A} \in \mathcal{Y}$, we have $\mathbf{A} \in \mathcal{K}$ if and only if $\mathbf{U} \not\leq \mathbf{A}$ for all $\mathbf{U} \in \mathbf{O}(\mathcal{K})$.*

The universal quasivariety $\mathcal{Y}$ just provides a frame of reference, e.g., lattices or all algebras of the given type.

PROOF. First note that $\mathcal{K}$ is the set of all finite models in $\mathcal{Y}$ of a set Σ of first-order sentences. Indeed, for each $k > 0$ we can rewrite

$$|\mathbf{A}| = k \implies \mathbf{A} \cong \mathbf{B}_1 \text{ OR } \cdots \text{ OR } \mathbf{A} \cong \mathbf{B}_m$$

as a first-order sentence, where $\mathbf{B}_1, \ldots, \mathbf{B}_m$ are the k-element algebras in $\mathcal{K}$.

Let $\mathcal{Q} = \mathbb{SPU}(\mathcal{K})$ and let $\mathcal{Q}_{\text{fin}}$ denote the finite members of $\mathcal{Q}$. Clearly $\mathcal{K} \subseteq \mathcal{Q}_{\text{fin}}$. For the reverse containment, let $\mathbf{C} \in \mathcal{Q}_{\text{fin}}$. Write the algebra $\mathbf{C}$ as a subdirect product of $\mathcal{Q}$-subdirectly irreducible algebras, $\mathbf{C} \leq \mathbf{C}_1 \times \cdots \times \mathbf{C}_n$. Each $\mathbf{C}_j$ is a finite algebra in $\mathbb{U}(\mathcal{K})$, and hence in $\mathcal{K}$ by the preceding observation. Since $\mathcal{K}$ is closed under $\mathbb{S}$ and $\mathbb{P}$ by assumption, then $\mathbf{C} \in \mathcal{K}$, as desired.

For the second part, let $\mathcal{R}$ denote the quasivariety generated by all finite algebras in $\mathcal{Y}$, and write $\mathcal{R} = \bigcup \mathcal{R}_n$ with $\mathcal{R}_1 \subseteq \mathcal{R}_2 \subseteq \cdots$ and each $\mathcal{R}_n$ a finitely generated quasivariety. In particular, each $\mathcal{R}_n$ is locally finite, and the theory from Section 3.1 applies.

Consider the quasivarieties $\mathcal{Q} \cap \mathcal{R}_n$. By Corollary 3.9(1), there is a set $\mathcal{O}_n$ of finite quasicritical algebras such that for every finite algebra $\mathbf{D} \in \mathcal{R}_n$, we have $\mathbf{D} \in \mathcal{Q} \cap \mathcal{R}_n$ if and only if $\mathbf{U} \not\leq \mathbf{D}$ for all $\mathbf{U} \in \mathcal{O}_n$. So let $\mathbf{O}(\mathcal{K}) = \bigcup \mathcal{O}_n$.

If $\mathbf{A}$ is a finite algebra in $\mathcal{Y}$, then $\mathbf{A} \in \mathcal{R}_{n_0}$ for some n_0. If $\mathbf{A} \notin \mathcal{K}$, i.e., $\mathbf{A} \notin \mathcal{Q}$ by (2), then $\mathbf{U} \leq \mathbf{A}$ for some $\mathbf{U} \in \mathcal{O}_{n_0} \subseteq \mathbf{O}(\mathcal{K})$. On the other hand, if $\mathbf{V} \leq \mathbf{A}$ for some $\mathbf{V} \in \mathbf{O}(\mathcal{K})$, then again by the construction in Section 3.1 the elements of $\mathbf{V}$ represent a failure of some quasi-equation holding in $\mathcal{Q}$, whence $\mathbf{A} \notin \mathcal{Q}$ and $\mathbf{A} \notin \mathcal{K}$. $\square$

An easy way to produce pseudoquasivarieties is to reverse engineer the characterization.

THEOREM 3.15. *Let $\mathcal{Y}$ be a quasivariety of algebras of finite type. Let $\mathcal{E}$ be a collection of finite $\mathcal{Y}$-subdirectly irreducible algebras. Then the class $\mathcal{K}$ of all finite algebras $\mathbf{A}$ in $\mathcal{Y}$ such that $\mathbf{U} \not\leq \mathbf{A}$ for all $\mathbf{U} \in \mathcal{E}$ is closed under $\mathbb{S}$ and $\mathbb{P}_{\text{fin}}$, and hence a pseudoquasivariety.*

For a concrete example, let E be a collection of prime power integers, and let $\mathcal{E}$ be the collection of all cyclic groups $\mathbf{Z}_q$ for $q \in E$. Then $\mathcal{K}$ consists of all finite abelian groups with no element of order q for any $q \in E$. The quasivariety $\mathcal{Q} = \mathbb{SPU}(\mathcal{K})$ is determined by the laws $p^k x \approx 0 \rightarrow p^{k-1} x \approx 0$ for all $q = p^k \in E$. Note that $\mathcal{Q}$ contains all torsion-free abelian groups. (See also Section 2.5.)

But examples from Chapter 5 show that the condition that the algebras in $\mathcal{E}$ be $\mathcal{Y}$-subdirectly irreducible is not necessary for the conclusion.

Following Adaricheva, Hyndman, Lempp, and Nation [16], we define the class $\mathcal{D}$ of *interval dismantlable* lattices by

(1) $\mathbf{1} \in \mathcal{D}$,
(2) if $\mathbf{L}$ is a finite lattice and there exists a join prime element $p \in L$ such that $\uparrow p \in \mathcal{D}$ and $\downarrow \kappa(p) \in \mathcal{D}$, where $\kappa(p) = \bigvee \{x \in L : x \not\geq p\}$, then $\mathbf{L} \in \mathcal{D}$,
(3) only these lattices are in $\mathcal{D}$.

Thus $\mathbf{L}$ is interval dismantlable if it can be partitioned into an ideal and a filter, each of which can be partitioned into an ideal and a filter, etc., until you reach 1-element lattices. It is straightforward that $\mathcal{D}$ is closed under taking sublattices and finite direct products.

In order to work with these lattices, we note that the following are equivalent for a finite lattice $\mathbf{L}$.

(1) $L = I \dot\cup F$ for some disjoint proper ideal I and filter F.
(2) $\mathbf{L}$ contains a nonzero join prime element.
(3) $\mathbf{L}$ contains a non-one meet prime element.
(4) There is a surjective homomorphism $h : \mathbf{L} \rightarrow \mathbf{2}$.
(5) Some generating set X for $\mathbf{L}$ can be split into two disjoint nonempty subsets, $X = Y \dot\cup Z$, such that $\bigwedge Y \not\leq \bigvee Z$.
(6) Every generating set X for $\mathbf{L}$ can be split into two disjoint nonempty subsets, $X = Y \dot\cup Z$, such that $\bigwedge Y \not\leq \bigvee Z$.

So if $\mathbf{L}$ contains no join prime element, then it is not interval dismantlable. If $\mathbf{L}$ contains no join prime element, but every proper sublattice does, then it is minimally interval non-dismantlable. If $\mathbf{L}$ contains an interval non-dismantlable sublattice, then $\mathbf{L}$ is interval non-dismantlable.

Note that it follows from (2) and (3) that every finite meet semidistributive or join semidistributive lattice is interval dismantlable. The atoms of a finite meet semidistributive lattice are join prime; dually, the coatoms of a finite join semidistributive lattice are meet prime.

In view of conditions (5) and (6) above, let us say that a subset X of a lattice $\mathbf{L}$ is *divisible* if it can be divided into two disjoint nonempty subsets Y and Z such that $\bigwedge Y \not\leq \bigvee Z$; else X is *indivisible*.

With respect to part (1) of Theorem 3.14, we can find a set of quasi-equations that determine $\mathcal{D}$. For each $n \geq 3$, let $X_n = \{x_1, \ldots, x_n\}$ be a set of n variables. Consider the quasi-equations

$$(\delta_n) \qquad \underset{\varnothing \subset Y \subset X_n}{\&} \bigwedge Y \leq \bigvee (X_n \setminus Y) \rightarrow x_1 \approx x_2 \, .$$

For example, δ_3 is

$$x \leq y \vee z \, \& \, y \leq x \vee z \, \& \, z \leq x \vee y \, \& \, x \wedge y \leq z \, \& \, x \wedge z \leq y \, \& \, y \wedge z \leq x$$
$$\to x \approx y.$$

Any indivisible subset A of a lattice $\mathbf{L}$ with $|A| \leq n$ satisfies the hypothesis of δ_n. On the other hand, by symmetry the conclusion could be replaced by $x_i \approx x_j$ for any $i \neq j$. Hence the quasi-equation δ_n expresses that $\mathbf{L}$ contains no indivisible subset of size k for $1 < k \leq n$. In particular, δ_n implies δ_{n-1}.

The main result of [16] is that the quasi-equations δ_n characterize interval dismantlable lattices.

THEOREM 3.16. *A finite lattice is interval dismantlable if and only if it satisfies δ_n for all $n \geq 3$, that is, the lattice contains no indivisible subset of more than one element.*

PROOF. First, assume that $\mathbf{L}$ is interval dismantlable. For every $n \geq 3$ and $\mathbf{a} \in L^n$, we want to show that δ_n holds under the substitution $x_i \mapsto a_i$. If $a_1 = a_2$, then the conclusion of δ_n holds. If $a_1 \neq a_2$, then the sublattice $\mathbf{S} = \mathrm{Sg}(a_1, \ldots, a_n)$ is nontrivial and interval dismantlable, and hence $\mathbf{S}$ has a decomposition $S = I \dot\cup F$ into a proper ideal and filter. Let $Y = \{a_i : i \in F\}$ and $Z = \{a_j : a_j \in I\}$. Then $\bigwedge Y \in F$ and $\bigvee Z \in I$, whence $\bigwedge Y \not\leq \bigvee Z$, so that the corresponding inclusion in the hypothesis of δ_n fails. Thus δ_n holds for every substitution.

Conversely, let us show that every finite lattice that satisfies all δ_n is interval dismantlable. We do so by induction on $|L|$. To begin, the 1-element lattice satisfies every δ_n and is in $\mathcal{D}$. So consider a finite lattice $\mathbf{L}$ with $|L| > 1$. Choose a generating set $X = \{a_1, a_2, \ldots, a_k\}$ for $\mathbf{L}$ with $a_1 \neq a_2$. Since $\mathbf{L}$ satisfies δ_k and the conclusion fails, there is a nontrivial splitting $X = Y \dot\cup Z$ with $\bigwedge Y \not\leq \bigvee Z$. This splits $\mathbf{L}$ into a proper ideal and filter, $L = I \dot\cup F$, and each of these is a smaller lattice that satisfies δ_n for all n. By induction, both I and F are interval dismantlable, and so $\mathbf{L}$ is also. $\square$

Any class of finite lattices closed under sublattices can be characterized by the exclusion of its minimal non-members. Minimal interval non-dismantlable lattices include $\mathbf{M}_3$, the four examples in Figure 3.1, and the duals of the two that are not self-dual. The lattices in the figure all fail δ_4. By varying the lattices in the corners of the bottom example, one can construct at least twenty more. But to show that the pseudoquasivariety $\mathcal{D}$ is not finitely based, we need an infinite sequence of minimal interval non-dismantlable lattices, such that any finite collection of the quasi-equations δ_j is satisfied in at least one of them. The next theorem provides this by generalizing the middle example.

THEOREM 3.17. *There is a sequence of minimal interval non-dismant-lable lattices $\mathbf{K}_n$ ($n \geq 4$) such that each $\mathbf{K}_n$ satisfies δ_j for $3 \leq j < n$, but fails δ_n.*

PROOF. For $n \geq 4$, we construct a lattice $\mathbf{K}_n$ as follows. The carrier set is $n \times (n-2) = \{(i,j) : 0 \leq i < n \text{ and } 0 \leq j < n-2\}$, with the order given

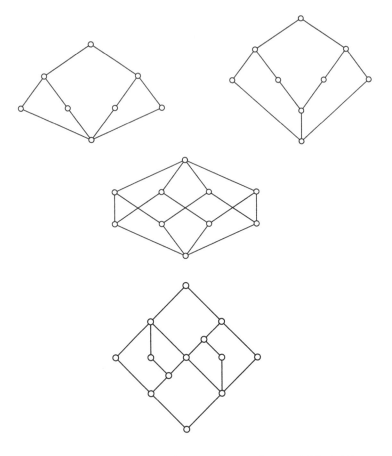

FIGURE 3.1. Four minimal interval non-dismantlable lattices.

by $(i, j) \leq (k, \ell)$ if $j \leq \ell$ and either $0 \leq k - i \leq \ell - j$ or $n + k - i \leq \ell - j$, plus a top element T and bottom element B. Thus we are thinking of the first coordinates modulo n, as if wrapped around a cylinder. The covers in the middle of the lattice are $(i, j) \prec (i, j + 1)$ and $(i, j) \prec (i + 1 \bmod n, j + 1)$ where $0 \leq i < n$ and $0 \leq j < n - 3$. The middle portion of the lattice $\mathbf{K}_5$ is illustrated in Figure 3.2.

For a generating set, we can take $X = \{(i, 0) : i < n\}$. This has the property that any pair of distinct elements of X meets to B, while the join of any $n - 1$ is T. Thus $\mathbf{K}_n$ fails δ_n and is interval non-dismantlable.

In view of the circular symmetry, we may consider the maximal sublattices not containing the generator $(0, 0)$. These are easily seen to be $S_0 = K_n \setminus \{(0, j) : j < n - 2\}$ and $\mathbf{T}_0 = K_n \setminus \{(j, j) : j < n - 2\}$. Both these sublattices are interval dismantlable. For we have $S_0 = {\uparrow}(1, 0) \cup {\downarrow}(n - 1, n - 3)$, with the filter being dually isomorphic to the lattice $\mathrm{Co}(\mathbf{n} - \mathbf{2})$ of convex subsets of an $n - 2$ element chain, and hence meet semidistributive, and the ideal being isomorphic to $\mathrm{Co}(\mathbf{n} - \mathbf{2})$ and hence join semidistributive. Likewise $T_0 = \uparrow(n-1, 0) \,\dot\cup\, \downarrow(n-2, n-3)$, with the filter being meet semidistributive and the ideal being join semidistributive.

To see that $\mathbf{K}_n$ satisfies δ_j for $3 \leq j < n$, consider an arbitrary generating set X for $\mathbf{K}_n$. For each k with $0 \leq k < n$, the set $\mathbf{S}_k = K_n \backslash \{(k, \ell) : \ell < n-2\}$ is a proper sublattice of $\mathbf{K}_n$. Hence $X \not\subseteq S_k$, i.e., X contains an element of the form (k, ℓ) for each $k < n$. Thus $|X| \geq n$. So every subset of K_n with fewer than n elements generates a proper sublattice, which is interval dismantlable. Therefore $\mathbf{K}_n$ satisfies δ_j for $j < n$. □

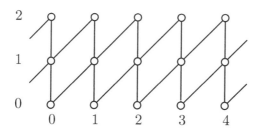

FIGURE 3.2. Middle portion of the lattice $\mathbf{K}_5$; add top and bottom elements for the whole lattice.

The original notion of dismantlability is that a finite lattice is *dismantlable* if it can be reduced to a 1-element lattice by successively removing doubly irreducible elements. These lattices were characterized independently by Ajtai [22] and Kelly and Rival [109], as those lattices not containing an n-crown for $n \geq 3$. Dismantlable lattices do not form a pseudoquasivariety, as they are not closed under finite direct products.

More generally, we can define a *sublattice dismantling* of a lattice to be a partition of the lattice into two nonempty, complementary sublattices. A finite lattice is said to be *sublattice dismantlable* if it can be reduced to 1-element lattices by successive sublattice dismantlings. Clearly, both the original dismantlable lattices and interval dismantlable lattices are sublattice dismantlable, and this class does form a pseudoquasivariety. It would be interesting to characterize sublattice dismantlable lattices.

CHAPTER 4

Analyzing $L_q(\mathcal{K})$

4.1. Algorithms

To further investigate the structure of $L_q(\mathcal{K})$, where $\mathcal{K}$ is a locally finite quasivariety of finite type, we want algorithms to determine

(1) the quasicritical algebras $\mathbf{T}$ in $\mathcal{K}$,
(2) the order on join irreducible quasivarieties, i.e., when $\langle \mathbf{T} \rangle \leq \langle \mathbf{S} \rangle$,
(3) the join dependencies, i.e., when $\langle \mathbf{T} \rangle \leq \langle \mathbf{S}_1 \rangle \vee \cdots \vee \langle \mathbf{S}_n \rangle$ nontrivially.

Let us show how these problems can be answered *locally*.

That is, given a finite algebra $\mathbf{T} \in \mathcal{K}$, we can determine which among $\mathbf{T}$ and its subalgebras are quasicritical, and the embedding order on those. Moreover, from the subdirect decompositions of $\mathbf{T}$ we can find all the minimal nontrivial join covers $\langle \mathbf{T} \rangle \leq \langle \mathbf{S}_1 \rangle \vee \cdots \vee \langle \mathbf{S}_n \rangle$ of $\langle \mathbf{T} \rangle$ in $L_q(\mathcal{K})$. On the other hand, the quasivariety $\mathcal{K}$ may have infinitely many other quasicritical algebras $\mathbf{U}$, including some with $\langle \mathbf{U} \rangle \leq \langle \mathbf{T} \rangle$ or $\langle \mathbf{T} \rangle \leq \langle \mathbf{U} \rangle$, and there may be no way to find these except by an infinite exhaustive search. There are no algorithms for the *global* structure of $L_q(\mathcal{K})$.

It is useful to record an elementary lemma that is used often. (A stronger version is Theorem 2.15.)

LEMMA 4.1. *Every finite algebra is either quasicritical or a subdirect product of quasicritical subalgebras.*

PROOF. If a finite algebra $\mathbf{T}$ is not quasicritical, then it is a subdirect product of proper subalgebras, say $\mathbf{T} \leq \mathbf{S}_1 \times \cdots \times \mathbf{S}_k$. By induction, each $\mathbf{S}_j$ is a subdirect product of its quasicritical subalgebras. □

For an element c and a finite subset A of a lattice $\mathbf{L}$, we say that A is a *join cover* of c if $c \leq \bigvee A$. The join cover is *nontrivial* if $c \not\leq a$ for all $a \in A$. Recall that for subsets A, B of $\mathbf{L}$, we say that A *refines* B, or B *refines to* A, if for every $a \in A$ there exists $b \in B$ with $a \leq b$. This is written $A \ll B$. The join cover $c \leq \bigvee A$ is *minimal* if whenever $c \leq \bigvee B$ and $B \ll A$, then $A \subseteq B$. Thus a minimal join cover consists of an antichain of join irreducible

© Springer International Publishing AG, part of Springer Nature 2018
J. Hyndman, J. B. Nation, *The Lattice of Subquasivarieties of a Locally Finite Quasivariety*, CMS Books in Mathematics,
https://doi.org/10.1007/978-3-319-78235-5_4

elements, such that no element of A can be omitted or replaced by a set of smaller elements.

LEMMA 4.2. *Let $\mathcal{K}$ be a locally finite quasivariety of finite type. If $\mathbf{T}$ is a finite algebra in $\mathcal{K}$, then every nontrivial join cover of $\langle \mathbf{T} \rangle$ in the quasivariety lattice $L_q(\mathcal{K})$ refines to one of finitely many nontrivial join covers $\langle \mathbf{T} \rangle \leq \langle \mathbf{R}_1 \rangle \vee \cdots \vee \langle \mathbf{R}_k \rangle$ with each $\mathbf{R}_j$ a finite quasicritical algebra in $\mathbb{H}(\mathbf{T})$. The nontrivial join covers of $\langle \mathbf{T} \rangle$ come from subdirect decompositions of $\mathbf{T}$.*

PROOF. Recall that every quasivariety in $L_q(\mathcal{K})$ is a join of completely join irreducible quasivarieties, and that these are of the form $\langle \mathbf{S} \rangle$ with $\mathbf{S}$ a finite quasicritical algebra in $\mathcal{K}$. If $\mathbf{T}$ is finite, then $\langle \mathbf{T} \rangle$ is compact in $L_q(\mathcal{K})$. Combining these facts, we see that every nontrivial cover of $\langle \mathbf{T} \rangle$ refines to one of the form $\langle \mathbf{T} \rangle \leq \langle \mathbf{S}_1 \rangle \vee \cdots \vee \langle \mathbf{S}_m \rangle$ with each $\mathbf{S}_j$ quasicritical. Then $\mathbf{T} \leq \mathbf{R}_1 \times \cdots \times \mathbf{R}_n$ with each $\mathbf{R}_i$ in some $\langle \mathbf{S}_j \rangle$, and we can assume that this representation is irredundant and that no $\mathbf{R}_i$ can be replaced by a set of its subalgebras. (There may be more than one $\mathbf{R}_i$ for a given $\mathbf{S}_j$, but they can all be taken to be quasicritical.) Since the cover is assumed to be nontrivial, this corresponds to a proper subdirect decomposition of $\mathbf{T}$ into quasicritical algebras in $\mathcal{K}$. There are only finitely many such, and they can be read off from the lattice of $\mathcal{K}$-congruences of $\mathbf{T}$, labeled by the isomorphism type of the factors $\mathbf{T}/\theta$.

The subdirect decomposition gives $\langle \mathbf{T} \rangle \leq \langle \mathbf{R}_1 \rangle \vee \cdots \vee \langle \mathbf{R}_n \rangle$. Taking only the quasivarieties $\langle \mathbf{R}_i \rangle$ that are maximal among those on this list, and renumbering, yields $\langle \mathbf{T} \rangle \leq \langle \mathbf{R}_1 \rangle \vee \cdots \vee \langle \mathbf{R}_k \rangle$. $\qquad\square$

Now we consider the order relation on join irreducible quasivarieties.

THEOREM 4.3. *Let $\mathcal{K}$ be a locally finite quasivariety of finite type. Let $\mathbf{S}$ and $\mathbf{T}$ be finite algebras in $\mathcal{K}$. If $\langle \mathbf{T} \rangle \leq \langle \mathbf{S} \rangle$ in $L_q(\mathcal{K})$, then either*

(1) *$\mathbf{T} \leq \mathbf{S}$ or*
(2) *there are quasicritical algebras $\mathbf{R}_1, \ldots, \mathbf{R}_m$ $(m \geq 1)$ with each $\mathbf{R}_j$ in $\mathbb{H}(\mathbf{T}) \cap \mathcal{K}$ such that*

$$\langle \mathbf{T} \rangle \leq \langle \mathbf{R}_1 \rangle \vee \cdots \vee \langle \mathbf{R}_m \rangle \leq \langle \mathbf{S} \rangle$$

in $L_q(\mathcal{K})$.

There may be infinitely many quasicritical algebras $\mathbf{S}$ of the first type. The second type corresponds to subdirect decompositions of $\mathbf{T}$ where the factors are all in $\langle \mathbf{S} \rangle$, so there are only finitely many candidates for the algebras $\mathbf{R}_1, \ldots, \mathbf{R}_m$.

PROOF. Again we write $\mathbf{T}$ as a subalgebra of a direct product of algebras in $\langle \mathbf{S} \rangle$, say $\mathbf{T} \leq \mathbf{R}_1 \times \cdots \times \mathbf{R}_m$ minimally with each $\mathbf{R}_i$ in $\langle \mathbf{S} \rangle$. If some projection is an embedding, then $m = 1$ and we have the first option. Otherwise, this is a proper subdirect decomposition of $\mathbf{T}$, and we have the second option. These can be read off from the labeled $\mathcal{K}$-congruence lattice of $\mathbf{T}$.

Of particular interest is the case when the join in (2) represents a trivial cover, so that $\langle \mathbf{T} \rangle < \langle \mathbf{R} \rangle \leq \langle \mathbf{S} \rangle$ with $\mathbf{R} \in \mathbb{H}(\mathbf{T})$ and $\mathbf{T} \leq \mathbf{R}^m$. $\qquad \square$

Of course, if $\mathbf{T}$ is k-generated and $\mathbf{T} < \mathbf{S}$, then there is a subalgebra $\mathbf{U} \leq \mathbf{S}$ with at most $k+1$ generators such that $\mathbf{T} < \mathbf{U} \leq \mathbf{S}$, and because of local finiteness there are only finitely many candidates for $\mathbf{U}$. However, it may happen that no such $\mathbf{U}$ is quasicritical, and we are primarily interested in the embedding order on quasicritical algebras. There the situation is more complicated. In Proposition 4.17 and Figure 4.3 of Section 4.5 we will see an example of a quasicritical algebra $\mathbf{Z}_1$ that is covered by infinitely many quasicritical algebras $\mathbf{Z}_n$ ($n \geq 2$) in the set of quasicriticals ordered by subalgebra inclusion. Thus we cannot expect to find a direct method to determine all the quasicritical algebras covering a given algebra $\mathbf{T}$ in the embedding order.

On the other hand, the fact that $\mathbf{T} \prec \mathbf{S}$ in the embedding order need not mean that $\langle \mathbf{T} \rangle$ is covered by $\langle \mathbf{S} \rangle$ as quasivarieties, and indeed the quasivarieties $\langle \mathbf{Z}_n \rangle$ in Section 4.5 form a chain (Figure 4.4). We do not know whether it is possible for a finitely generated quasivariety $\langle \mathbf{T} \rangle$ to have infinitely many upper covers in $\mathrm{L_q}(\mathcal{K})$ when $\mathcal{K}$ is locally finite and has finite type. This is Problem 5 in Chapter 8.

In Section 5.3 we will see how part (2) of the previous theorem can be useful. From the congruence lattice of the algebra $\mathbf{R}_1$ in Figure 5.8, we see that $\mathbf{R}_1$ has a subdirect representation $\mathbf{R}_1 \leq \mathbf{T}_1 \times \mathbf{R}_2$. However, $\mathbf{T}_1 \leq \mathbf{R}_2$, so $\mathbf{R}_1 \leq \mathbf{R}_2^2$ and hence $\langle \mathbf{R}_1 \rangle \leq \langle \mathbf{R}_2 \rangle$. Similarly we obtain $\mathbf{R}_1 \leq \mathbf{T}_1 \times \mathbf{R}_3 \leq \mathbf{R}_3^2$ and so $\langle \mathbf{R}_1 \rangle \leq \langle \mathbf{R}_3 \rangle$. But more subtly, $\mathbf{R}_1 \leq \mathbf{T}_2 \times \mathcal{F}_1$ with both $\mathbf{T}_2, \mathcal{F}_1 \leq \mathbf{R}_7$, and hence $\langle \mathbf{R}_1 \rangle \leq \langle \mathbf{R}_7 \rangle$. This last inclusion is not immediately apparent just from inspecting $\mathrm{Con}_{\mathcal{K}} \mathbf{R}_1$.

Now let us consider how we can analyze the structure surrounding the quasivariety $\langle \mathbf{T} \rangle$ generated by a finite algebra in $\mathcal{K}$. Suppose we are given $\mathbf{T}$, and that we know

- the quasicriticals $\mathbf{S}_1, \ldots, \mathbf{S}_k$ that are properly in $\mathbb{S}(\mathbf{T})$,
- the quasicriticals $\mathbf{U}_1, \ldots, \mathbf{U}_\ell$ that are properly in $\mathbb{HS}(\mathbf{T}) \setminus \mathbb{S}(\mathbf{T})$,
- the order and dependencies on $\langle \mathbf{S}_1 \rangle, \ldots, \langle \mathbf{S}_k \rangle, \langle \mathbf{U}_1 \rangle, \ldots, \langle \mathbf{U}_\ell \rangle$, that is, when one quasivariety is below the join of others.

We may assume inductively that all this information is known for the algebras properly in $\mathbb{HS}(\mathbf{T})$.

Now we start on $\langle \mathbf{T} \rangle$ itself. The first question is: *Is $\mathbf{T}$ quasicritical?* This can be reliably answered by the Universal Algebra Calculator. If the answer is NO, then $\langle \mathbf{T} \rangle$ is in the quasivariety generated by its quasicritical subalgebras $\mathbf{S}_1, \ldots, \mathbf{S}_k$, for which we already know the information, and we are done.

If the answer is YES, meaning that $\mathbf{T}$ is quasicritical, add $\mathbf{T}$ to the list of quasicriticals: $\mathbf{S}_1, \ldots, \mathbf{S}_k, \mathbf{T}$ and the order relations $\langle \mathbf{S}_j \rangle < \langle \mathbf{T} \rangle$. It remains to add the information on order and dependencies contained in $\mathrm{Con}_{\mathcal{K}} \mathbf{T}$.

If $\mathbf{T}$ is $\mathcal{K}$-subdirectly irreducible, then $\langle \mathbf{T} \rangle$ is completely join prime in $\mathrm{L_q}(\mathcal{K})$. In this case $\langle \mathbf{T} \rangle \leq \langle \mathbf{R} \rangle$ holds only when $\mathbf{T} \leq \mathbf{R}$, by Corollaries 3.5

and 3.6. Moreover, we never have $\mathbf{T} < \mathbf{S}_i$ or $\mathbf{T} < \mathbf{U}_j$ for the algebras in our list, since those are properly in $\mathbb{HS}(\mathbf{T})$. So there is no more information to add.

But if $\mathbf{T}$ is subdirectly reducible in $\mathcal{K}$, then we consider in turn each irredundant proper decomposition $\Delta = \theta_1 \wedge \cdots \wedge \theta_m$ in $\mathrm{Con}_{\mathcal{K}} \mathbf{T}$ with all factors $\mathbf{T}/\theta$ quasicritical, and hence either $\mathbf{T}/\theta \cong \mathbf{S}_i$ or $\mathbf{T}/\theta \cong \mathbf{U}_j$ for one of the quasicriticals in our list. (Note that the congruences θ need not be meet irreducible, but if $\mathbf{T}/\theta$ is not quasicritical, then that θ can be replaced by a meet of larger congruences where the factors are quasicritical.)

Let $\mathbf{R}_k \cong \mathbf{T}/\theta_k$ for $1 \leq k \leq m$. Then we have $\langle \mathbf{T} \rangle \leq \langle \mathbf{R}_1 \rangle \vee \cdots \vee \langle \mathbf{R}_m \rangle$. Since $\mathbf{T}$ is quasicritical, at least one $\mathbf{R}_i$ must come from $\mathbf{U}_1, \ldots, \mathbf{U}_\ell$.

Again there are two possibilities. If for some $\mathbf{R}_p$ (necessarily one of the $\mathbf{U}_j$'s) we have $\mathbf{R}_j \in \langle \mathbf{R}_p \rangle$ for all j, then $\langle \mathbf{T} \rangle < \langle \mathbf{R}_p \rangle$. On the other hand, if there is no such $\mathbf{R}_p$, then $\langle \mathbf{T} \rangle \leq \langle \mathbf{R}_1 \rangle \vee \cdots \vee \langle \mathbf{R}_m \rangle$ is a nontrivial join cover in $L_q(\mathcal{K})$. Examples of both types, inclusions and nontrivial dependencies, will appear in the lattice of subquasivarieties of $\mathcal{M}$ described in Chapter 5.

This analysis can be summarized as follows. Given a finite algebra $\mathbf{T} \in \mathcal{K}$, we look at the irredundant subdirect decompositions of $\mathbf{T}$ into quasicritical algebras in $\mathcal{K}$. These give us relations of the forms

(1) $\langle \mathbf{T} \rangle \leq \langle \mathbf{R}_1 \rangle \vee \cdots \vee \langle \mathbf{R}_k \rangle$ with each $\mathbf{R}_j \in \mathbb{H}(\mathbf{T}) \cap \mathcal{K}$,
(2) $\langle \mathbf{T} \rangle \leq \langle \mathbf{V}_1 \rangle \vee \cdots \vee \langle \mathbf{V}_m \rangle \leq \langle \mathbf{S} \rangle$ with each $\mathbf{V}_j \in \mathbb{H}(\mathbf{T}) \cap \mathcal{K}$.

The second option is closely related to the first. Because $\mathbf{T}$ is finite, there are only finitely many relations of these types.

The other type of order relation that holds among the compact elements of $L_q(\mathcal{K})$ are those $\langle \mathbf{T} \rangle \leq \langle \mathbf{S} \rangle$ that arise from embeddings $\mathbf{T} \leq \mathbf{S}$. For these, we basically have to construct the larger algebra. Offhand, there could be infinitely many such extensions that are minimal in the sense that $\mathbf{T} < \mathbf{S}$, but there is no quasicritical algebra with $\mathbf{T} < \mathbf{U} < \mathbf{S}$. However, the Universal Algebra Calculator can be used to find the subalgebras of a given algebra $\mathbf{S}$.

4.2. Fermentability

In this section we describe some properties of subquasivariety lattices $L_q(\mathcal{K})$ that hold when $\mathcal{K}$ is a locally finite quasivariety of finite type, but not necessarily in the lattice $L_q(\mathcal{Q})$ for an arbitrary quasivariety. Let us begin with a little lattice theory, taken from Chapter II of [69] but based on work of Day [49], Jónsson [94], and McKenzie [115].

We begin by considering lower boundedness for finitely generated lattices. A lattice homomorphism $h : \mathbf{K} \to \mathbf{L}$ is a *lower bounded homomorphism* if for every element $a \in L$, the inverse image $h^{-1}(\uparrow a) = \{w \in K : h(w) \geq a\}$ is either empty or has a least element. A finitely generated lattice $\mathbf{L}$ is a *lower bounded lattice* if it is a lower bounded homomorphic image of a finitely generated free lattice, i.e., if there is a surjective lower bounded homomorphism $h : \mathbf{FL}(X) \twoheadrightarrow \mathbf{L}$ for some finite set X. Lower bounded lattices initially arose in the context of sublattices of free lattices and projective lattices.

Let $\mathbf{L}$ be a lattice. We say that a subset $A \subseteq L$ has the *minimal join cover refinement property* if for every $a \in A$, every nontrivial finite cover $a \leq \bigvee C$ can be refined to one of finitely many minimal finite covers $M \ll C$, so that $a \leq \bigvee M$, all of which satisfy $M \subseteq A$. We say that $\mathbf{L}$ has the minimal join cover refinement property if the property holds with $A = L$.

Let $\mathrm{D}_0(\mathbf{L})$ be the set of all join prime elements of $\mathbf{L}$, i.e., the set of all elements that have no nontrivial join cover. Given $\mathrm{D}_k(\mathbf{L})$, define $\mathrm{D}_{k+1}(\mathbf{L})$ to be the set of all $p \in L$ such that every nontrivial join cover of p refines to a join cover contained in $\mathrm{D}_k(\mathbf{L})$, i.e., $p \leq \bigvee C$ nontrivially implies there exists $B \ll C$ with $p \leq \bigvee B$ and $B \subseteq \mathrm{D}_k(\mathbf{L})$. Note that if $p \in A$ where $A \subseteq L$ has the minimal join cover refinement property, then $p \in \mathrm{D}_{k+1}(\mathbf{L})$ if and only if every minimal nontrivial join cover of p is contained in $\mathrm{D}_k(\mathbf{L})$.

Clearly $\mathrm{D}_k(\mathbf{L}) \subseteq \mathrm{D}_{k+1}(\mathbf{L})$ for all $k \in \omega$. Let $\mathrm{D}(\mathbf{L}) = \bigcup_{k \in \omega} \mathrm{D}_k(\mathbf{L})$.

The basic theorem on lower bounded lattices can be stated thusly.

THEOREM 4.4. *For a finitely generated lattice $\mathbf{L}$, the following are equivalent.*

(1) *$\mathbf{L}$ is a lower bounded lattice, i.e., there is a surjective lower bounded homomorphism $h : \mathbf{FL}(X) \twoheadrightarrow \mathbf{L}$ for some finite set X.*
(2) *For every finitely generated lattice $\mathbf{K}$, every homomorphism $h : \mathbf{K} \to \mathbf{L}$ is lower bounded.*
(3) *$\mathrm{D}(\mathbf{L}) = \mathbf{L}$.*

Moreover, if $\mathbf{L}$ satisfies these properties, then it is join semidistributive and $\mathbf{L}$ has the minimal join cover refinement property.

Examples of lower bounded lattices include any finitely generated sublattice of a free lattice [94], and the subalgebra lattice of any finite semilattice [10, 70]. For an extension of lower boundedness to lattices that may not be finitely generated, see Adaricheva and Gorbunov [14].

With these tools in hand, we return to lattices of subquasivarieties. The restriction to locally finite quasivarieties of finite type brings a strong restriction, due to Adaricheva, Dziobiak, and Gorbunov [12], based on Adaricheva [10].

THEOREM 4.5. *Let $\mathcal{K}$ be a locally finite quasivariety of finite type. If $\mathrm{L}_q(\mathcal{K})$ is finite, then it is a lower bounded lattice.*

It is this fact that we will generalize in Theorem 4.7 below. On the other hand, there are finite lower bounded lattices that do not support any equaclosure operator (e.g., those in Figure A.1 in the Appendix). A lattice that does not support an equaclosure operator cannot be represented as $\mathrm{L}_q(\mathcal{Q})$ for any quasivariety $\mathcal{Q}$.

The lattice $\mathrm{Co}(4)$ of convex subsets of a 4-element chain does not support an equaclosure operator, so it cannot be represented as a subquasivariety lattice. As $\mathrm{Co}(4)$ is not lower bounded, neither is any lattice containing it as a sublattice. However, the *leaf lattice* $\mathbf{L}_4$ in Figure 4.1 is isomorphic to

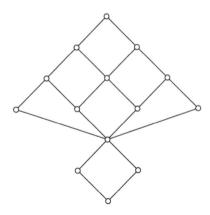

FIGURE 4.1. Leaf lattice $\mathbf{L}_4$ that is isomorphic to $L_q(\mathcal{Q})$ for a quasivariety $\mathcal{Q}$, but not to $L_q(\mathcal{K})$ for any locally finite quasivariety $\mathcal{K}$ of finite type.

$L_q(\mathcal{Q})$ for a quasivariety $\mathcal{Q}$ constructed in [19], though it contains $\mathrm{Co}(\mathbf{4})$. By Theorem 4.5, there is no locally finite quasivariety $\mathcal{K}$ of finite type such that $\mathbf{L}_4 \cong L_q(\mathcal{K})$. The similar lattice $\mathbf{1} + \mathrm{Co}(\mathbf{4})$ is isomorphic to a lattice of H-closed algebraic subsets $S_p(\mathbf{L}, H)$ for a pair $(\mathbf{L}, H)$, also from [19], but we don't know whether $\mathbf{1} + \mathrm{Co}(\mathbf{4})$ can be represented as a lattice of subquasivarieties. (See Section A.1 of the Appendix for the representation of subquasivariety lattices by lattices of algebraic subsets.)

In the next lemma, the *length* of a finite lattice $\mathbf{L}$ is the maximum length of a chain in $\mathbf{L}$, where a chain $c_0 < c_1 < \cdots < c_n$ has length n.

LEMMA 4.6. *If the length of* $\mathrm{Con}_{\mathcal{K}} \mathbf{T}$ *is* k, *then the quasivariety* $\langle \mathbf{T} \rangle$ *is in* $D_{k-1}(L_q(\mathcal{K}))$.

PROOF. If $\mathrm{Con}_{\mathcal{K}} \mathbf{T}$ has length 1, then $\mathbf{T}$ is simple, and hence $\langle \mathbf{T} \rangle$ is join prime and in $D_0(L_q(\mathcal{K}))$. If the length is $k \geq 2$, then either $\mathbf{T}$ is $\mathcal{K}$-subdirectly irreducible, in which case $\langle \mathbf{T} \rangle$ is join prime in $L_q(\mathcal{K})$, or else the minimal nontrivial join covers of $\langle \mathbf{T} \rangle$ correspond to proper subdirect decompositions of $\mathbf{T}$, with each factor having a shorter $\mathcal{K}$-congruence lattice, and hence being in $D_{k-2}(L_q(\mathcal{K}))$ by induction. $\square$

A lattice is *fermentable* if it is $\bigvee$-generated by some set A of join irreducible elements such that

(a) A has the minimal join cover refinement property,

(b) $A \subseteq D(\mathbf{L})$.

These lattices were introduced in Wehrung [167], generalizing Pudlák and Tůma [146].

To see that $L_q(\mathcal{K})$ is fermentable when $\mathcal{K}$ is a locally finite quasivariety of finite type, we take A to be the collection of all quasivarieties $\langle \mathbf{T} \rangle$ with $\mathbf{T}$ quasicritical. By Lemma 4.2, every join cover of such a $\langle \mathbf{T} \rangle$ refines to a cover $\langle \mathbf{T} \rangle \leq \langle \mathbf{R}_1 \rangle \vee \cdots \vee \langle \mathbf{R}_k \rangle$ with each $\mathbf{R}_j$ a quasicritical algebra in $\mathbb{H}(\mathbf{T})$.

By Lemma 4.6 we have $\langle \mathbf{T} \rangle \in D(L_q(\mathcal{K}))$. Combining these facts yields the following result.

THEOREM 4.7. *If $\mathcal{K}$ is a locally finite quasivariety of finite type, then the lattice of subquasivarieties $L_q(\mathcal{K})$ is fermentable.*

Fermentability carries strong consequences for the structure of $L_q(\mathcal{K})$. In particular, we have the following result of Wehrung [167], found independently by Semenova [154].

THEOREM 4.8. *Every fermentable lattice can be embedded into a direct product of finite lower bounded lattices.*

Let LB_{fin} denote the class of finite lower bounded lattices, and let $\mathbb{Q}(LB_{\mathrm{fin}})$ be the quasivariety it generates.

COROLLARY 4.9. *If $\mathcal{K}$ is a locally finite quasivariety of finite type, then the lattice $L_q(\mathcal{K})$ is in $\mathbb{Q}(LB_{\mathrm{fin}})$.*

It is easy to find quasi-equations that hold in $\mathbb{Q}(LB_{\mathrm{fin}})$. Remember that every sublattice of a lower bounded lattice is lower bounded. Let $\mathbf{M}$ be a finite subdirectly irreducible lattice that is not lower bounded, with (a, b) a critical pair for $\mathbf{M}$. Then $\mathbf{M}$ cannot be embedded into any lower bounded lattice, and so $\mathbb{Q}(LB_{\mathrm{fin}})$ satisfies

$$\mathrm{diag}(\mathbf{M}) \to a \approx b.$$

In particular, we could take $\mathbf{M}$ to also be join semidistributive, for example, $\mathbf{M} = \mathrm{Co}(4)$, the lattice of convex subsets of a 4-element chain. Every lattice $L_q(\mathcal{K})$, with $\mathcal{K}$ a locally finite quasivariety of finite type, satisfies these quasi-equations.

On the other hand, every lattice Sub $\mathbf{S}$ of subsemilattices of a finite semilattice is isomorphic to $L_q(\mathcal{K})$ for a quasivariety of one-element structures with finitely many relations, by Theorem A.5, and these lattices satisfy no nontrivial lattice *equation* [71].

When we compare the situation for subquasivariety lattices without the locally finite, finite type restriction, a different picture emerges. The next theorem combines results of Adaricheva, Gorbunov, and Tumanov [15, 78, 164]. Let QL denote the class of all subquasivariety lattices $L_q(\mathcal{Q})$, and let $\langle \mathrm{SD}_\vee \rangle$ denote the quasivariety of all join semidistributive lattices.

THEOREM 4.10. $\mathbb{Q}(QL) = \langle \mathrm{SD}_\vee \rangle$, *that is, subquasivariety lattices generate $\langle \mathrm{SD}_\vee \rangle$.*

PROOF. Any join semidistributive lattice can be embedded into an ultraproduct of finite join semidistributive lattices. Thus the quasivariety $\langle \mathrm{SD}_\vee \rangle$ is generated by its finite members.

Moreover, every finite join semidistributive lattice $\mathbf{K}$ can be embedded into a lattice of algebraic subsets $S_p(\mathbf{L})$ for some algebraic lattice $\mathbf{L}$. Note that if $\mathbf{K}$ is not lower bounded, then $\mathbf{L}$ will necessarily be infinite.

Finally, $S_p(\mathbf{L})$ is isomorphic to $L_q(\mathcal{Q})$ for a quasivariety of one-element structures, by Theorem A.5. Thus a finite join semidistributive lattice embeds into $L_q(\mathcal{Q})$ for some quasivariety $\mathcal{Q}$.   □

There is a 4-generated join semidistributive lattice that cannot be embedded into any lower continuous, join semidistributive lattice; in particular, it cannot be embedded into any subquasivariety lattice $L_q(\mathcal{Q})$. This is Example 4.4.15 in Gorbunov [77], reproduced as Example 4-1.32 in [20]. It follows from the proof of Theorem 4.10 that *the class of lattices that can be embedded into subquasivariety lattices is not first-order axiomatizable* [106].

Lemma 4.6 has another, more local consequence for calculations in $L_q(\mathcal{K})$ when $\mathcal{K}$ is locally finite and finite type.

In general, if a lattice $\mathbf{L}$ is algebraic and join-generated by its compact join irreducible elements, then $\mathbf{L}$ is determined by the order on $J(\mathbf{L})$ and the join covers $p \leq \bigvee Q$ with $p \in J(\mathbf{L})$ and Q a finite subset of $J(\mathbf{L})$. If in addition $J(\mathbf{L})$ has the minimal join cover refinement property, then we only need the minimal nontrivial join covers, i.e., those where Q is minimal in the sense of refinement ($\ll$), of which there are only finitely many. (Recall that A *refines* B, written $A \ll B$, if for every $a \in A$ there exists $b \in B$ with $a \leq b$.) And if also $J(\mathbf{L}) \subseteq D(\mathbf{L})$, then we need only the minimal nontrivial join covers $p \leq \bigvee Q$ with $\bigvee Q$ also minimal in the order $\leq$ of $\mathbf{L}$. That is, if $p \leq \bigvee Q$ and $p \leq \bigvee R$ and $\bigvee Q \leq \bigvee R$, we keep only $p \leq \bigvee Q$. (See Section II.4 of Freese, Ježek, and Nation [69] or Section 9 of Adaricheva, Nation, and Rand [21].) Thus, if a lattice $\mathbf{L}$ satisfies

(1) $\mathbf{L}$ is algebraic and dually algebraic,
(2) $J(\mathbf{L})$ has the minimal join cover refinement property,
(3) $J(\mathbf{L}) \subseteq D(\mathbf{L})$,

then $\mathbf{L}$ is determined by its so-called *E-basis*, consisting of the order on $J(\mathbf{L})$ and the finite nontrivial join covers $p \leq \bigvee Q$ with p and each $q \in Q$ join irreducible, that are nonrefinable and have $\bigvee Q$ minimal in $\mathbf{L}$.

All this applies to $L_q(\mathcal{K})$ when $\mathcal{K}$ is a locally finite quasivariety of finite type.

COROLLARY 4.11. *If $\mathcal{K}$ is a locally finite quasivariety of finite type, then $L_q(\mathcal{K})$ is determined by*

(1) *the order* $\langle \mathbf{S} \rangle \leq \langle \mathbf{T} \rangle$ *on quasivarieties generated by quasicritical algebras in $\mathcal{K}$,*
(2) *the nonrefinable join covers* $\langle \mathbf{S} \rangle \leq \langle \mathbf{T}_1 \rangle \vee \cdots \vee \langle \mathbf{T}_m \rangle$ *with $\mathbf{S}$ and each $\mathbf{T}_i$ quasicritical, and $\bigvee_i \langle \mathbf{T}_i \rangle$ minimal in $L_q(\mathcal{K})$.*

This corollary can be useful when computing $L_q(\mathcal{K})$ for a specific quasivariety $\mathcal{K}$.

4.3. Equational Quasivarieties

Before tackling a concrete example of a quasivariety lattice $L_q(\mathcal{K})$, let us record in this section some observations that will be useful when we do so.

We consider the question: *which subquasivarieties are equational?* That is, which subquasivarieties of a quasivariety $\mathcal{K}$ are determined, relative to $\mathcal{K}$, by sets of equations?

The equational quasivarieties play an important role in the structure of the lattice $L_q(\mathcal{K})$, as they determine the so-called *equaclosure operator*. This was introduced in Dziobiak [55], with additional properties found in [13] and [19]; see Section A.4 of the Appendix for a description. But also, we will see equational quasivarieties in the lattice $L_q(\mathcal{M})$ used to illustrate the general method in Chapter 5.

Here are three ways in which equational quasivarieties arise, in increasing generality.

(1) If $\mathbf{F}$ is a finite $\mathcal{K}$-free algebra, then within $\mathcal{K}$ the quasi-equation $\mathrm{diag}(\mathbf{F}) \to s \approx t$ is equivalent to $s \approx t$, since the premises always hold. Hence $\langle \mathrm{diag}(\mathbf{F}) \to s \approx t \rangle$ is equal to $\langle s \approx t \rangle$.

(2) If $\mathcal{V}$ is a variety contained in $\mathcal{K}$, and $\mathbf{F}$ is a $\mathcal{V}$-free algebra, then the quasivariety $\mathcal{V} \wedge \langle \mathrm{diag}(\mathbf{F}) \to s \approx t \rangle$ is equal to $\mathcal{V} \wedge \langle s \approx t \rangle$, which is equational.

(3) For any quasi-equation $\beta: (\&_i s_i \approx t_i) \to u \approx v$, if $\mathcal{W}$ is the variety determined by the premises $(\&_i s_i \approx t_i)$ of β, then $\mathcal{W} \wedge \langle \beta \rangle$ is equational, being equal to $\mathcal{W} \wedge \langle u \approx v \rangle$.

The first type can give us meet irreducible equational quasivarieties, and the meet of equational quasivarieties is of course equational. The latter two indicate ways in which meet reducible equational quasivarieties can be formed.

This is at least a start on this aspect of the problem, and again it is illustrated in Chapter 5.

4.4. The Atoms of $L_q(\mathcal{K})$

Now we turn to describing the quasivarieties that are atoms of $L_q(\mathcal{K})$ when $\mathcal{K}$ is locally finite. In describing atoms, we must deal with algebras and relational structures separately.

The basic result for atoms in quasivariety lattices of algebras is due to Ahmad Shafaat [159].

THEOREM 4.12. *Let $\mathcal{K}$ be a locally finite quasivariety of algebras of finite type. Then $\mathcal{B}$ is an atom in $L_q(\mathcal{K})$ if and only if $\mathcal{B} = \langle \mathbf{T} \rangle$ for a finite algebra $\mathbf{T}$ such that $|T| > 1$ and*

$$(\dagger) \qquad \textit{if } \mathbf{U} \leq \mathbf{T}^2 \textit{ and } |U| > 1, \textit{ then } \mathbf{T} \leq \mathbf{U}.$$

Note that, since $\mathbf{T} \leq \mathbf{T}^2$ *via* the diagonal embedding, an algebra $\mathbf{T}$ satisfying ($\dagger$) has no proper subalgebra with more than one element, and hence is quasicritical.

PROOF. Let $\mathcal{B}$ be an atom of $L_q(\mathcal{K})$. Then $\mathcal{B}$ is completely join irreducible, so it is generated by a finite quasicritical algebra $\mathbf{T}$. Since $\langle \mathbf{T} \rangle_*$ is the trivial quasivariety, $\mathbf{T}$ can have no proper subalgebra with more than one element. Theorem 2.28 then applies to yield ($\dagger$).

Conversely, consider an algebra $\mathbf{T}$ satisfying (†). As noted above, this implies that $\mathbf{T}$ is quasicritical with no non-singleton proper subalgebra. Corollary 2.29 then says that $\langle\mathbf{T}\rangle_*$ is the quasivariety generated by the proper subalgebras of $\mathbf{T}$, which in this case is the trivial quasivariety. □

Theorem 4.12 is easy to apply, since it is straightforward to check the condition (†) for a given finite algebra $\mathbf{T}$.

COROLLARY 4.13. *Let $\mathcal{K}$ be a locally finite quasivariety of algebras of finite type. If $\mathbf{T} \in \mathcal{K}$ with $|T| = 2$, then $\langle\mathbf{T}\rangle$ is an atom of $L_q(\mathcal{K})$.*

PROOF. Let $\mathbf{T}$ be a 2-element algebra in $\mathcal{K}$, say $T = \{a, b\}$. Clearly every proper subalgebra of $\mathbf{T}$ has one element. Consider a subalgebra $\mathbf{S} \leq \mathbf{T}^2$ with $|S| > 1$. If (a, a) and (b, b) are both in S, then they form a subalgebra isomorphic to $\mathbf{T}$. If neither is in S, then $|S| = 2$, so the projection maps must be one-to-one, and hence isomorphisms.

Thus we may assume say $(b, b) \in S$ and $(a, a) \notin S$. It follows that b is an idempotent element under all the operations of $\mathbf{T}$. Now at least one of (a, b), (b, a) is in S, say the former. Then $\{(a, b), (b, b)\}$ forms a subalgebra of $\mathbf{S}$ isomorphic to $\mathbf{T}$ *via* the first projection, as desired. □

Corollary 4.13 does not extend to 3-element algebras. In fact, the quasivariety generated by a 3-element algebra can be Q-universal, and hence have $2^{\aleph_0}$ subquasivarieties. See Adams and Dziobiak [2] and our Section 5.5.

COROLLARY 4.14. *Let $\mathcal{K}$ be a locally finite quasivariety of algebras of finite type. If $\langle\mathbf{T}\rangle$ is an atom of $L_q(\mathcal{K})$, then $\mathbf{T}$ is generated by at most 2 elements. Thus $L_q(\mathcal{K})$ has only finitely many atoms.*

Without local finiteness, $L_q(\mathcal{Q})$ can have infinitely many atoms, as we have seen for abelian groups in Section 2.5. See, e.g., Sapir [150] or Dziobiak, Ježek, and Maróti [58] for other examples.

Algebras that generate minimal subvarieties of a locally finite variety $\mathcal{V}$, i.e., the atoms of the lattice of subvarieties $L_v(\mathcal{V})$, are described in Kearnes and Szendrei [107]. An earlier survey by Szendrei [163] describes the minimal subvarieties of many classical varieties. Also, Bergman and McKenzie [28] proved that every locally finite, congruence modular, minimal variety is minimal as a quasivariety. They also gave examples to show that neither "local finiteness" nor "congruence modularity" can be dropped.

We say that a pure relational structure $\mathbf{R}$ is *full* if every possible relation hold in $\mathbf{R}$, i.e., $A^{\mathbf{R}} = R^n$ for each n-ary relation in the type. The characterization of atoms in lattices of quasivarieties of pure relational structures is due to Gorbunov and Tumanov [78].

THEOREM 4.15. *Let $\mathcal{K}$ be a locally finite quasivariety of pure relational structures of finite type. Then $\mathcal{C}$ is an atom in $L_q(\mathcal{K})$ if and only if $\mathcal{C} = \langle\mathbf{R}\rangle$ where either*

(1) *$|R| = 1$ and $\mathbf{R}$ is not full or*
(2) *$|R| = 2$ and $\mathbf{R}$ is full.*

PROOF. Again $\mathcal{C}$ is generated by a quasicritical structure $\mathbf{R}$. Let $\mathbf{P}_0$ denote the 1-element full structure in $\mathcal{K}$, and note $\mathbf{P}_0 = \prod \varnothing \neq \mathbf{R}$. If $|R| = 1$, then $\mathbf{R}$ is not full and $\{\mathbf{R}, \mathbf{P}_0\}$ is a subquasivariety of $\mathcal{K}$ contained in $\mathcal{C}$, whence equal to $\mathcal{C}$. If $|R| > 1$, then $|R| = 2$ as $\mathbf{T}$ has no nontrivial proper substructure. Let $R = \{x, y\}$. If some relation $A(x)$ or $B(y)$ or $C(x, y)$ failed to hold in $\mathbf{R}$, then either $\mathbf{R}$ or $\mathbf{R}^2$ would contain a nonfull 1-element substructure, which would generate a proper subquasivariety of $\mathcal{C}$, a contradiction. $\qquad\square$

Theorem 4.15 is illustrated in the quasivarieties of Chapter 7.

We leave the case of mixed algebraic structures, with both operations and relations, as a exercise for the reader.

Recall that a lattice is *biatomic* if whenever $p \succ 0$ and $p \leq a \vee b$, then there exist atoms $q \leq b$ and $r \leq c$ with $p \leq q \vee r$. Adaricheva and Gorbunov [13] showed that any atomic lattice supporting an equaclosure operator is biatomic; see Section A.4 of the Appendix. In particular, this applies to lattices of subquasivarieties $L_q(\mathcal{K})$.

4.5. The Variety $\mathcal{Z}$

Let us illustrate these methods by applying them to subquasivarieties of a variety $\mathcal{Z}$ of unary algebras. The variety $\mathcal{Z}$ is similar to, but simpler than, the variety $\mathcal{M}$ considered in Chapter 5, making it perhaps more appropriate for a first example.

Let $\mathcal{Z}$ be the variety of 2-unary algebras with operations f, g satisfying

$$f^2 x \approx f x \approx g f x \qquad g^2 x \approx g x \approx f g x.$$

Note that if an element a in an algebra in $\mathcal{Z}$ satisfies $fa = a$, then $ga = gfa = fa = a$. Similarly, $ga = a$ implies $fa = a$.

Within this variety, let $\mathbf{Z}_0$ denote the 1-element algebra, and consider the following algebras, which are drawn in Figure 4.2.

(1) Let $\mathbf{X}$ denote the algebra with universe $\{a, b\}$ and $fa = ga = fb = gb = b$.

(2) Let $\mathbf{Y}_1$ denote the algebra with universe $\{a, b\}$ and $fa = ga = a$, $fb = gb = b$.

(3) Let $\mathbf{Y}_2$ denote the algebra with universe $\{a, b, c, d\}$ and $fa = fb = c$, $ga = gb = d$ with c, d fixed under both operations.

(4) Let $\mathbf{Y}_3$ denote the algebra with universe $\{a, b, c, d, e, h\}$ and $fa = gb = d$, $fb = fc = e$, $ga = gc = h$ with d, e, h fixed under both operations.

(5) Let $\mathbf{Y}_4$ denote the algebra with universe $\{a, b, c, d, e\}$ and $fa = fb = gc = d$, $ga = gb = fc = e$ with d, e fixed under both operations.

(6) Let $\mathbf{Z}_1$ be the free algebra on one generator, with elements x, fx, gx.

(7) For $n \geq 2$ let $\mathbf{Z}_n$ be the algebra generated by $a_1, \ldots, a_n$ with $fa_1 = ga_2$, $fa_2 = ga_3, \ldots, fa_n = ga_1$.

Each of the algebras listed above is quasicritical. The algebras **X** and **Z**₁ are the 1-generated quasicriticals in $\mathcal{Z}$, while $\mathbf{Y}_1$, $\mathbf{Y}_2$, and $\mathbf{Z}_2$ are the 2-generated quasicriticals, and $\mathbf{Y}_3$, $\mathbf{Y}_4$, and $\mathbf{Z}_3$ are the 3-generated quasicriticals. All the quasicritical algebras in $\mathcal{Z}$ except **X**, $\mathbf{Y}_1$, and $\mathbf{Z}_1$ itself are obtained by gluing together copies of $\mathbf{Z}_1$ in various ways. We have singled out the algebras $\mathbf{Z}_n$, obtained by gluing $fa_j = ga_{j+1}$ cyclically, to illustrate some particular phenomena in $L_q(\mathcal{Z})$. The algebras $\mathbf{Y}_2$, $\mathbf{Y}_3$, and $\mathbf{Y}_4$ are other variations of the gluing construction.

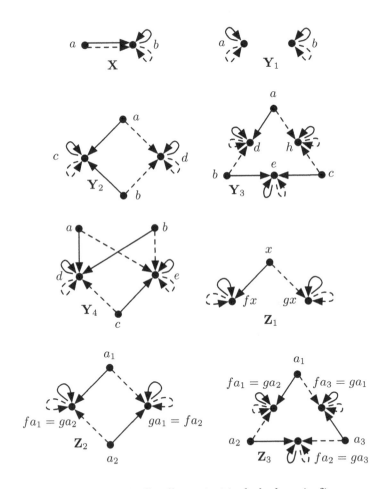

FIGURE 4.2. Small quasicritical algebras in $\mathcal{Z}$

The embedding order on these quasicritical algebras is given in Figure 4.3. Clearly $\mathbf{Z}_1 < \mathbf{Z}_n$ for $n \geq 2$, while $\{\mathbf{Z}_n : n \geq 2\}$ forms an antichain with respect to embedding.

Note that $\mathbf{Y}_1 < \mathbf{Z}_1$. Corollary 4.13 and straightforward calculations using the criterion of Lemma 2.28 yield the following covering relations.

LEMMA 4.16. *In* $L_q(\mathcal{Z})$ *we have* $\langle \mathbf{Z}_0 \rangle \prec \langle \mathbf{Y}_1 \rangle \prec \langle \mathbf{Z}_1 \rangle$.

PROPOSITION 4.17. *In the ordered set of quasicritical algebras of* $\mathcal{Z}$ *ordered by embedding,* $\mathbf{Z}_n$ *covers* $\mathbf{Z}_1$: *there is no quasicritical algebra* $\mathbf{S} \in \mathcal{Z}$ *with* $\mathbf{Z}_1 < \mathbf{S} < \mathbf{Z}_n$.

PROOF. For $n \geq 2$, the subalgebra of $\mathbf{Z}_n$ obtained by removing a generator a_i, that is $\mathbf{Z}_n \setminus \{a_i\}$, is in the quasivariety generated by $\mathbf{Z}_1$, and hence so is every proper subalgebra. But the only quasicritical algebras in $\langle \mathbf{Z}_1 \rangle$ are $\mathbf{Y}_1$ and $\mathbf{Z}_1$ by Lemma 4.16. □

This validates one of the claims of Theorem 4.3 that a quasicritical algebra can have infinitely many upper covers in the set of quasicritical algebras ordered by embedding.

The next proposition collects facts about the quasivarieties generated by some of these algebras, based on the algorithms earlier in this chapter and routine calculations. Note for the calculations that because $\mathcal{Z}$ is a variety, $\mathrm{Con}_{\mathcal{Z}} \mathbf{T} = \mathrm{Con}\,\mathbf{T}$ for any $\mathbf{T} \in \mathcal{Z}$.

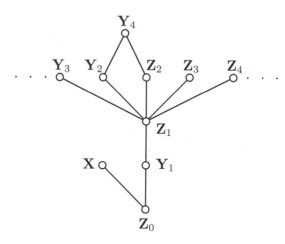

FIGURE 4.3. Small quasicritical algebras in $\mathcal{Z}$ ordered by embedding

PROPOSITION 4.18. *The quasivarieties generated by these small quasicritical algebras have the following properties.*

(1) $\mathbf{X}$ *is a 2-element simple algebra, so* $\langle \mathbf{X} \rangle$ *is a join prime atom of* $L_q(\mathcal{Z})$.

(2) $\mathbf{Y}_1$ *is a 2-element simple algebra, so* $\langle \mathbf{Y}_1 \rangle$ *is a join prime atom of* $L_q(\mathcal{Z})$.

(3) $\mathbf{Y}_2$ *is subdirectly irreducible, so* $\langle \mathbf{Y}_2 \rangle$ *is join prime in* $L_q(\mathcal{Z})$.

(4) $\mathbf{Y}_3 \leq \mathbf{Y}_2^2$, *so* $\langle \mathbf{Y}_3 \rangle \leq \langle \mathbf{Y}_2 \rangle$. *However,* $\mathbf{Y}_3$ *has a unique minimal nonzero reflection congruence (with* $\mathbf{Y}_3/\alpha \cong \mathbf{Z}_2$), *so* $\langle \mathbf{Y}_3 \rangle$ *is join prime.*

(5) $\mathbf{Y}_4 \leq \mathbf{X} \times \mathbf{Z}_2$, *so* $\langle \mathbf{Y}_4 \rangle \leq \langle \mathbf{X} \rangle \vee \langle \mathbf{Y}_2 \rangle$. *Also* $\langle \mathbf{Y}_2 \rangle \vee \langle \mathbf{Z}_2 \rangle < \langle \mathbf{Y}_4 \rangle$ *due to the embeddings.*

Because an atom of $L_q(\mathcal{Z})$ must be $\langle \mathbf{T} \rangle$ for a quasicritical algebra $\mathbf{T}$ with at most 2 generators, the only atoms are $\langle \mathbf{X} \rangle$ and $\langle \mathbf{Y}_1 \rangle$.

Now we turn to the quasivarieties $\langle \mathbf{Z}_n \rangle$.

PROPOSITION 4.19. *The quasivarieties* $\langle \mathbf{Z_n} \rangle$ *have these properties.*

(1) *Each* $\langle \mathbf{Z}_n \rangle$ *for* $n \geq 1$ *is join prime in* $L_q(\mathcal{Z})$.

(2) $\langle \mathbf{Z}_0 \rangle \prec \langle \mathbf{Y}_1 \rangle \prec \langle \mathbf{Z}_1 \rangle$ *in* $L_q(\mathcal{Z})$.

(3) $\langle \mathbf{Z}_1 \rangle < \langle \mathbf{Z}_n \rangle$ *for all* $n \geq 2$.

(4) $\langle \mathbf{Z}_n \rangle \leq \langle \mathbf{Z}_{n-1} \rangle$ *for all* $n > 2$.

(5) *Thus* $\langle \mathbf{Z}_0 \rangle < \langle \mathbf{Z}_1 \rangle < \cdots < \langle \mathbf{Z}_4 \rangle < \langle \mathbf{Z}_3 \rangle < \langle \mathbf{Z}_2 \rangle$.

(6) $\bigcap_{n \geq 2} \langle \mathbf{Z}_n \rangle = \langle \mathbf{Z}_1 \rangle$, *and hence* $\langle \mathbf{Z}_1 \rangle$ *is not finitely based.*

These results are illustrated in the segment of $L_q(\mathcal{Z})$ in Figure 4.4.

FIGURE 4.4. Segment of $L_q(\mathcal{Z})$. Ticks indicate known covers.

PROOF. Each $\langle \mathbf{Z}_n \rangle$ is join prime because the reflection into $\langle \mathbf{X} \rangle$ is the unique minimal nonzero reflection congruence on $\mathbf{Z}_n$. Indeed, the congruence $\theta = \bigvee_{1 \leq i \leq n} \mathrm{Cg}(fa_i, ga_i)$ is the smallest nonzero characteristic congruence, and reflection congruences are characteristic by Lemma 2.23. Thus (1) holds, while (2) is Lemma 4.16. Part (3) is due to the embedding $\mathbf{Z}_1 < \mathbf{Z}_n$.

For (4), on an algebra $\mathbf{Z}_n$ with $n > 2$, define the congruences ψ_i for $1 \le i \le n$ to have one block $[fa_i, a_i, ga_i]$ and the rest singletons. Then $\mathbf{Z}_n/\psi_i \cong \mathbf{Z}_{n-1}$ and $\psi_1 \cap \psi_2 = \Delta$, so that $\mathbf{Z}_n \le \mathbf{Z}_{n-1}^2$. Hence $\langle \mathbf{Z}_n \rangle \le \langle \mathbf{Z}_{n-1} \rangle$. Part (5) just summarizes this.

To prove (6), let $\mathbf{T} \in \bigcap_{n \ge 2} \langle \mathbf{Z}_n \rangle$ be an algebra with more than one element. We may assume that $\mathbf{T}$ is a finitely generated quasicritical algebra, say with k generators. If $k = 1$, then $\mathbf{T} \cong \mathbf{X}$ or $\mathbf{T} \cong \mathbf{Z}_1$; the former is impossible because $\mathbf{X} \notin \langle \mathbf{Z}_2 \rangle$ (as it satisfies the semi-splitting quasi-equation for $\mathbf{Z}_2$, which is $\mathrm{diag}(\mathbf{Z}_2) \to fx \approx gx$), and the latter is what we are trying to prove. So assume $k > 1$. Then in particular $\mathbf{T} \in \langle \mathbf{Z}_{k+1} \rangle$, so that $\mathbf{T} \le \mathbf{Z}_{k+1}^m$ for some m. But then the projection maps $\pi_j : \mathbf{T} \to \mathbf{Z}_{k+1}$ are onto k-generated subalgebras of $\mathbf{Z}_{k+1}$, and as observed in the proof of Proposition 4.17, the proper subalgebras of $\mathbf{Z}_{k+1}$ are all in $\langle \mathbf{Z}_1 \rangle$. Thus $\mathbf{T} \in \langle \mathbf{Z}_1 \rangle$, as desired. It follows that $\langle \mathbf{Z}_1 \rangle$ is not dually compact in $\mathrm{L_q}(\mathcal{Z})$, i.e., not finitely based. $\square$

Note that part (6) of the preceding theorem shows that even a quasivariety of finite height in $\mathrm{L_q}(\mathcal{Z})$ need not be finitely based.

It was observed in part (4) of Proposition 4.18 that $\langle \mathbf{Y}_3 \rangle \le \langle \mathbf{Y}_2 \rangle$. Gluing copies of $\mathbf{Z}_1$ in a cycle, but using $fa_j = ga_{j+1}$ or $fa_j = fa_{j+1}$ or $ga_j = fa_{j+1}$ or $ga_j = ga_{j+1}$ in different patterns, yields various other quasicritical algebras. In this way one can find infinitely many quasicritical algebras in $\langle \mathbf{Y}_2 \rangle$, similar to what was done for $\langle \mathbf{Z}_2 \rangle$ with the algebras $\mathbf{Z}_n$. We leave the details as an exercise for the reader.

Thus we see how the algorithms can be used to analyze the structure of $\mathrm{L_q}(\mathcal{Z})$. The variety $\mathcal{M}$ of Chapter 5 turns out to be a good deal more complicated, but in interesting ways.

4.6. Synopsis

Here is a synopsis of the plot so far. Let $\mathcal{K}$ be a locally finite quasivariety of finite type. The lattice $\mathrm{L_q}(\mathcal{K})$ of subquasivarieties is algebraic, dually algebraic, join semidistributive, and fermentable. It is also atomic, admits an equaclosure operator, and satisfies the Jónsson-Kiefer property. These latter properties are discussed in Sections A.3 and A.4 of the Appendix.

The compact quasivarieties are those generated by a finite set of finite algebras. Each compact quasivariety has a canonical join representation as a join of finitely many completely join irreducibles. The completely join irreducible quasivarieties are exactly those generated by a single finite quasicritical algebra $\mathbf{T}$. Note that such a join irreducible quasivariety $\langle \mathbf{T} \rangle$ may contain infinitely many other quasivarieties $\langle \mathbf{S} \rangle$ with $\mathbf{S}$ quasicritical. The quasivariety $\langle \mathbf{T} \rangle$ is join prime in $\mathrm{L_q}(\mathcal{K})$ if and only if $\mathbf{T}$ has a unique minimal nonzero reflection $\mathcal{K}$-congruence.

For each finite algebra $\mathbf{T}$ and atom $\alpha \succ \Delta$ of $\mathrm{Con}_{\mathcal{K}} \mathbf{T}$, choose a pair a, $b \in \mathbf{T}$ such that $\alpha = \mathrm{Cg}_{\mathcal{K}}(a, b)$. Then define the quasi-equation

$$\varepsilon_{\mathbf{T}, \alpha}: \quad \mathrm{diag}(\mathbf{T}) \to t_a \approx t_b$$

where t_a, t_b are terms that evaluate to a, b, respectively, in $\mathbf{T}$. The basic result is that for any quasivariety $\mathcal{Q} \leq \mathcal{K}$, we have that $\mathbf{T} \notin \mathcal{Q}$ if and only if $\mathcal{Q}$ satisfies $\varepsilon_{\mathbf{T}, \alpha}$ for some α. The completely meet irreducible quasivarieties in $L_q(\mathcal{K})$ are all of the form $\langle \varepsilon_{\mathbf{T}, \alpha} \rangle$, but not conversely, and meet representations of finitely based varieties in terms of completely meet irreducibles need not be canonical.

Now, for a finite algebra $\mathbf{T} \in \mathcal{K}$, let

$$\mathcal{E}(\mathbf{T}) = \{\varepsilon_{\mathbf{T}, \alpha} : \alpha \succ \Delta\}$$

$$\lambda(\mathbf{T}) = \{\langle \varepsilon \rangle : \varepsilon \in \mathcal{E}(\mathbf{T})\}$$

$$\kappa(\mathbf{T}) = \{\mathcal{N} \in L_q(\mathcal{K}) : \mathcal{N} \text{ is maximal with respect to } \mathbf{T} \notin \mathcal{N}\}.$$

Then $\kappa(\mathbf{T}) \subseteq \lambda(\mathbf{T})$ since the quasivarieties in $\kappa(\mathbf{T})$ are completely meet irreducible.

However, within $\lambda(\mathbf{T})$ all sorts of things can happen. For distinct atoms α, $\beta \succ \Delta$ in $\mathrm{Con}_{\mathcal{K}} \mathbf{T}$, the quasivarieties $\langle \varepsilon_{\mathbf{T}, \alpha} \rangle$ and $\langle \varepsilon_{\mathbf{T}, \beta} \rangle$ could be distinct and incomparable. But it is also possible that $\langle \varepsilon_{\mathbf{T}, \alpha} \rangle = \langle \varepsilon_{\mathbf{T}, \beta} \rangle$, or that $\langle \varepsilon_{\mathbf{T}, \alpha} \rangle < \langle \varepsilon_{\mathbf{T}, \beta} \rangle$. Concrete examples of these phenomena will be given in Section 5.2. In addition, a quasivariety $\langle \varepsilon_{\mathbf{T}, \alpha} \rangle$ in $\lambda(\mathbf{T})$ can be meet reducible.

For any finite algebra $\mathbf{T} \in \mathcal{K}$ and atom α of $\mathrm{Con}_{\mathcal{K}} \mathbf{T}$ there exist finite sets $\mathbf{N}(\mathbf{T}, \alpha) \subseteq \mathbf{O}(\mathbf{T}, \alpha)$ that determine whether a quasivariety $\mathcal{Q} \leq \mathcal{K}$ satisfies $\mathcal{Q} \leq \langle \varepsilon_{\mathbf{T}, \alpha} \rangle$ by exclusion, with respect to subquasivarieties or subalgebras, respectively. That is, for an algebra $\mathbf{A} \in \mathcal{K}$, $\mathbf{A} \in \langle \varepsilon_{\mathbf{T}, \alpha} \rangle$ holds if and only if $\mathbf{U} \not\leq \mathbf{A}$ for all $\mathbf{U} \in \mathbf{O}(\mathbf{T}, \alpha)$, and for a quasivariety $\mathcal{Q} \leq \mathcal{K}$, we have $\mathcal{Q} \leq \langle \varepsilon_{\mathbf{T}, \alpha} \rangle$ if and only if $\mathbf{R} \notin \mathcal{Q}$ for all $\mathbf{R} \in \mathbf{N}(\mathbf{T}, \alpha)$. This information can be used to describe the order and dual dependence on the meet irreducible quasivarieties, and to characterize when a quasivariety is finitely based. For example, a subquasivariety $\mathcal{Q}$ of a locally finite quasivariety $\mathcal{K}$ of finite type is finitely based relative to $\mathcal{K}$ if and only if there exist finitely many finite quasicritical algebras $\mathbf{R}_1, \ldots, \mathbf{R}_m$ such that $\langle \mathbf{R}_i \rangle \not\leq \mathcal{Q}$ but $\langle \mathbf{R}_i \rangle_* \leq \mathcal{Q}$ for all i, and for any subquasivariety $\mathcal{W} \leq \mathcal{K}$, if $\mathcal{W} \not\leq \mathcal{Q}$ then $\mathbf{R}_j \in \mathcal{W}$ for some j. This is not true without the local finiteness assumption.

When $\mathbf{T} \in \mathcal{K}$ is finite, then $\langle \mathbf{T} \rangle$ has only finitely many nontrivial minimal join covers $\langle \mathbf{T} \rangle \leq \langle \mathbf{S}_1 \rangle \vee \cdots \vee \langle \mathbf{S}_n \rangle$ in $L_q(\mathcal{K})$, and these can be found by inspecting $\mathrm{Con}_{\mathcal{K}} \mathbf{T}$. It is this property that makes $L_q(\mathcal{K})$ fermentable.

CHAPTER 5

Unary Algebras with 2-Element Range

5.1. The Variety Generated by M

The previous sections have included various algorithms for working with locally finite quasivarieties of finite type. We will illustrate these algorithms by applying them to quasivarieties contained in the variety $\mathcal{M}$ generated by a particular 3-element algebra $\mathbf{M}$ described below. Conveniently, because the quasivariety for our example is the whole variety $\mathcal{M} = \mathbb{HSP}(\mathbf{M})$, we have $\mathrm{Con}_{\mathcal{M}} \mathbf{T} = \mathrm{Con}\, \mathbf{T}$ for all $\mathbf{T} \in \mathcal{M}$.

The type of $\mathcal{M}$ is $(1,1,0)$, i.e., it has two unary operations, denoted f and g, and a constant, denoted 0. The universe of $\mathbf{M}$ is $\{0,1,2\}$ and its operations are given in Table 5.1.

	f	g
0	0	0
1	0	1
2	1	0

TABLE 5.1. The operations of $\mathbf{M}$. Note that $\mathrm{Rows}(\mathbf{M})$, the rows of the operation table, is an order ideal in $\mathbf{2}^2$.

Unary algebras with a constant 0 and a 2-element range $\{0,1\}$ for their operations were studied in Casperson *et al.* [43]. There it was shown that if the rows of the operation table form an order ideal of $\mathbf{2}^n$, then the algebra has a finite basis for its quasi-equations. This applies to our algebra $\mathbf{M}$. In fact, Theorem 5.2 below shows that the 2-variable quasi-equations of $\mathbb{Q}(\mathbf{M})$ form a basis. (The basis is given in Proposition 5.24(1); later in the chapter, the algebra $\mathbf{M}$ is designated as $\mathbf{T}_3$.) Nonetheless, we shall see that $\mathbb{Q}(\mathbf{M})$ contains infinitely many finite quasicritical algebras, and it contains uncountably many subquasivarieties.

Finding an equational basis for the variety $\mathcal{M} = \mathbb{V}(\mathbf{M})$ is not hard.

© Springer International Publishing AG, part of Springer Nature 2018
J. Hyndman, J. B. Nation, *The Lattice of Subquasivarieties of a Locally Finite Quasivariety*, CMS Books in Mathematics,
https://doi.org/10.1007/978-3-319-78235-5_5

THEOREM 5.1. *The variety* $\mathcal{M}$ *is determined by the equations*

$$f^2 x \approx fgx \approx 0 \qquad g^2 x \approx gx \qquad gfx \approx fx \qquad f0 \approx g0 \approx 0.$$

PROOF. If $\mathcal{V}$ is any variety of unary algebras, perhaps with constants, then the diagram of the free algebra $\mathcal{F}_{\mathcal{V}}(2)$ forms an equational basis for $\mathcal{V}$. As observed early on by Birkhoff [31], if $\mathcal{V}$ is generated by a finite unary algebra, then $\mathcal{F}_{\mathcal{V}}(2)$ is finite, and thus $\mathcal{V}$ is finitely based. In our case, the free algebra $\mathcal{F}_{\mathcal{M}}(2)$ consists of two copies of $\mathcal{F}_{\mathcal{M}}(1)$ glued together at 0. Hence any 2-variable identity $s(x) \approx t(y)$ satisfied by $\mathcal{M}$ is equivalent to two 1-variable identities, $s(x) \approx 0$ and $t(x) \approx 0$. Thus the diagram of $\mathcal{F}_{\mathcal{M}}(1)$, given in the theorem statement, is a basis for $\mathcal{M}$.   □

The free algebra $\mathcal{F}_1 = \mathcal{F}_{\mathcal{M}}(1)$ generated by x, and its quasicritical homomorphic images, are illustrated in Figure 5.1. The quasicritical algebras in $\mathbb{HS}(\mathcal{F}_1)$ and $\mathbb{HS}(\mathcal{F}_2)$ are referred to as the *small quasicritical algebras*. These algebras are actually in $\mathbb{H}(\mathcal{F}_1)$ and $\mathbb{H}(\mathcal{F}_2)$, and can be described by the elements of the free algebra equated by homomorphisms. These descriptions are provided in Table 5.2. Although $\mathcal{F}_1$ is quasicritical, $\mathcal{F}_k$ is not when $k > 1$. The small quasicritical algebras in $\mathcal{M}$ are illustrated in Figures 5.1 and 5.2. In the figures, solid arrows represent the operation f, while dashed arrows indicate g. Note that $\mathbf{T}_3$ is isomorphic to $\mathbf{M}$. The 1-element algebra $\mathbf{T}_0$, though not quasicritical, is included in these figures for reference.

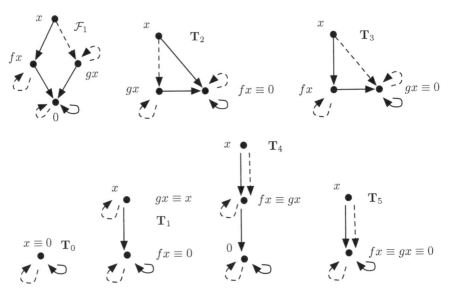

FIGURE 5.1. Homomorphic images of $\mathcal{F}_1$ that are quasicritical.

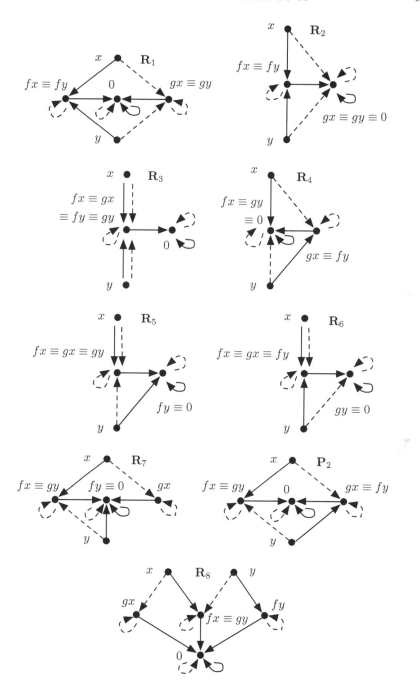

FIGURE 5.2. Homomorphic images of $\mathcal{F}_2$ that are quasicritical.

alg	# gen	# elts	equated free algebra elements
$\mathcal{F}_1$	1	4	
$\mathbf{T}_0$	1	1	$x \equiv 0$
$\mathbf{T}_1$	1	2	$x \equiv gx$ and $fx \equiv 0$
$\mathbf{T}_2$	1	3	$fx \equiv 0$
$\mathbf{T}_3$	1	3	$gx \equiv 0$
$\mathbf{T}_4$	1	3	$fx \equiv gx$
$\mathbf{T}_5$	1	2	$fx \equiv gx \equiv 0$
$\mathbf{P}_2$	2	5	$fx \equiv gy$ and $gx \equiv fy$
$\mathbf{R}_1$	2	5	$fx \equiv fy$ and $gx \equiv gy$
$\mathbf{R}_2$	2	4	$fx \equiv fy$ and $gx \equiv gy \equiv 0$
$\mathbf{R}_3$	2	4	$fx \equiv fy \equiv gx \equiv gy$
$\mathbf{R}_4$	2	4	$fx \equiv gy \equiv 0$ and $gx \equiv fy$
$\mathbf{R}_5$	2	4	$fx \equiv gx \equiv gy$ and $fy \equiv 0$
$\mathbf{R}_6$	2	4	$fx \equiv gx \equiv fy$ and $gy \equiv 0$
$\mathbf{R}_7$	2	5	$fx \equiv gy$ and $fy \equiv 0$
$\mathbf{R}_8$	2	6	$fx \equiv gy$

TABLE 5.2. Small quasicritical algebras as homomorphic images of $\mathcal{F}_1$ and $\mathcal{F}_2$.

In the next two sections we shall apply our algorithms to the small quasicritical algebras of $\mathcal{M}$. First, the 1-generated algebras will be considered, and then those that are 2-generated. (*All* the 1-generated algebras of $\mathcal{M}$ are quasicritical; not so the 2-generated algebras.) In Section 5.4, we show that $L_q(\mathcal{M})$ is uncountable. Section 5.5 proves the stronger fact that the quasivariety $\langle \mathbf{T}_3 \rangle$ is Q-universal.

5.2. Illustrating the Algorithms: 1-Generated Algebras

Let $\mathbf{T}$ be a small quasicritical algebra in $\mathcal{M}$. To analyze the structure of $L_q(\mathcal{M})$ surrounding $\langle \mathbf{T} \rangle$, we need the following information.

- A list of the small quasicritical algebras in $\mathcal{M}$, given in Table 5.2.
- A chart giving the embedding relations amongst the small quasi-criticals, given in Figure 5.3.
- The labeled congruence lattice of $\mathbf{T}$. (For the 1-generated algebras considered in this section, this information can be read off from the congruence lattice of the free algebra on one generator, given in Figure 5.4.) *Throughout, for u, $v \in \mathbf{T}$, we use the notation $u \equiv v$ to represent the congruence $\mathrm{Cg}(u,v)$.*
- Inductively, inclusions $\langle \mathbf{R} \rangle \leq \langle \mathbf{S} \rangle$ for quotient algebras $\mathbf{R} \cong \mathbf{T}/\varphi$, $\mathbf{S} \cong \mathbf{T}/\theta$. (In other words, when $\mathbf{R} \leq \mathbf{S}^n$ for some n.)
- A list of the reflection congruences of $\mathrm{Con}_{\mathcal{K}} \mathbf{T}$, obtained by applying Theorem 2.19 to the above information.

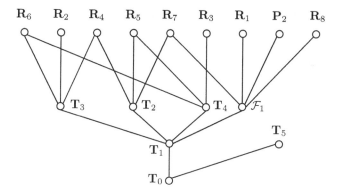

FIGURE 5.3. Small quasicritical algebra embeddings in $\mathcal{M}$

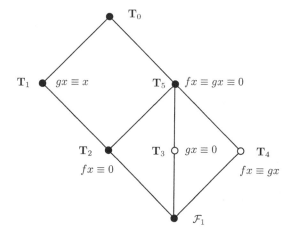

FIGURE 5.4. The congruence lattice of $\mathcal{F}_1$.

Given an algebra $\mathbf{R}$, the set of all quasivarieties maximal with respect to not being above $\langle \mathbf{R} \rangle$ is denoted $\kappa(\mathbf{R})$. Each quasivariety $\mathcal{U}$ in $\kappa(\mathbf{R})$ is completely meet irreducible with upper cover $\mathcal{U}^*$. Given an atom α in the $\mathcal{K}$-congruence lattice of $\mathbf{R}$, the quasivariety $\mathcal{U}(\mathbf{R}, \alpha)$ is defined by the quasi-equation $\varepsilon_{\mathbf{R},\alpha}$, i.e., $\mathrm{diag}(\mathbf{R}) \to s \approx t$ where $\alpha = \mathrm{Cg}(s, t)$. Recall that $\lambda(\mathbf{R})$ denotes the set of quasivarieties of the form $\mathcal{U}(\mathbf{R}, \alpha)$ with α an atom of the $\mathcal{K}$-congruence lattice. Further recall that $\kappa(\mathbf{R}) \subseteq \lambda(\mathbf{R})$, but this may be a proper containment. If $\mathcal{U}(\mathbf{R}, \alpha)$ *is in* $\kappa(\mathbf{R})$, *then we write* $\widehat{\mathcal{U}}(\mathbf{R}, \alpha)$ *to emphasize this fact.*

The format of the analysis in these sections is this: For each small quasi-critical algebra $\mathbf{T}$ in $\mathcal{M}$ we determine some or all of the following information.

(1) A list of some useful quasi-equations satisfies by $\mathbf{T}$, including a basis for $\langle \mathbf{T} \rangle$ if known.

(2) A list of the small quasicritical algebras in $\langle \mathbf{T} \rangle$.

(3) Some small quasicritical algebras such that $\langle \mathbf{S} \rangle$ is minimal with respect to $\langle \mathbf{S} \rangle \nleq \langle \mathbf{T} \rangle$, as in Theorem 3.4(2).

(4) Some small quasicritical algebras such that $\mathbf{S}$ is minimal with respect to $\mathbf{S} \notin \langle \mathbf{T} \rangle$. (So every proper subalgebra of $\mathbf{S}$ is in $\langle \mathbf{T} \rangle$; compare Theorem 3.4(1).)

(5) Description of the lower cover $\langle \mathbf{T} \rangle_*$ if known.

(6) A list of the reflection congruences of Con $\mathbf{T}$. (*These are indicated by solid dots in the figures.*)

(7) $\gamma^{\mathbf{T}}(\alpha)$ for the atoms α of Con $\mathbf{T}$.

(8) The diagram diag($\mathbf{T}$).

(9) The set of $\mathcal{E}(\mathbf{T})$ of quasi-equations $\varepsilon_{\mathbf{T},\alpha}$ for atoms α of Con $\mathbf{T}$. (Then $\lambda(\mathbf{T})$ is the set of all $\langle \varepsilon_{\mathbf{T},\alpha} \rangle$ with $\varepsilon_{\mathbf{T},\alpha} \in \mathcal{E}(\mathbf{T})$.)

(10) $\kappa(\mathbf{T})$ using Theorem 2.32.

(11) Quasi-equations for $\mathcal{U}(\mathbf{T}, \alpha)$ and/or $\widehat{\mathcal{U}}(\mathbf{T}, \alpha)$.

(12) Description of $\mathcal{U}(\mathbf{T}, \alpha)$ and/or $\widehat{\mathcal{U}}(\mathbf{T}, \alpha)$ by omitted subalgebras, i.e., $\mathbf{O}(\mathbf{T}, \alpha)$ from Theorem 3.4. (The sets $\mathbf{O}(\mathbf{T}, \alpha)$ and $\mathbf{N}(\mathbf{T}, \alpha)$ for small quasicritical algebras are given in Table 5.4.)

(13) A segment of $L_q(\mathcal{M})$ relating to $\mathbf{T}$.

A couple of comments are in order. These calculations are done in a rather different order than presented. We would probably start with (2): Given a pair of algebras, we can determine whether $\mathbf{S} \in \mathbb{SP}(\mathbf{T})$ from the labeled congruence lattice of $\mathbf{S}$ and the subalgebras of $\mathbf{T}$. Except near the bottom of $L_q(\mathcal{M})$, the quasivariety $\langle \mathbf{T} \rangle$ may contain infinitely many quasicriticals, so the list will necessarily be incomplete.

Then we would do the rather straightforward computations for (6)–(12). For an atom α of the congruence lattice, the reflection $\rho_{\langle \mathbf{T}/\alpha \rangle}$ is either α or Δ, depending on whether or not α is a reflection congruence. This information is implicit in (6). If $\mathbf{T}$ has a unique minimal nonzero reflection congruence, then $\langle \mathbf{T} \rangle$ is join prime; if not, we can find its minimal nontrivial join covers from the subdirect decompositions of $\mathbf{T}$.

With respect to part (5), we should admit that we do not really have a good method for describing $\langle \mathbf{T} \rangle_*$ in general, but we can do so for some of the small quasivarieties. Of course, we do know that $\langle \mathbf{T} \rangle_* = \langle \mathbf{T} \rangle \wedge \widehat{\mathcal{U}}(\mathbf{T}, \alpha)$ for some atom α, and sometimes Theorem 2.28 or Corollary 2.29 applies to show that $\langle \mathbf{T} \rangle_*$ is the quasivariety generated by the proper subalgebras of $\mathbf{T}$.

A summary of our investigation of the lower part of $L_q(\mathcal{M})$ is recorded in Table 5.3. The table gives some inclusions and nontrivial join covers in $L_q(\mathcal{M})$. This information is illustrated in the schematic diagram of $L_q(\mathcal{M})$ in Figure 5.5. The table includes only the inclusions that appear to be covers in the figure; the partial picture of $L_q(\mathcal{M})$ in Figure 5.5 also uses the embeddings given in Figure 5.3.

Fact	Reason
$\langle \mathbf{T}_0 \rangle \leq \langle \mathbf{T}_j \rangle$ for $j = 1, 5$	$\mathbf{T}_0 \leq \mathbf{T}_j$ for $j = 1, 5$
$\langle \mathbf{T}_1 \rangle \leq \langle \mathbf{T}_2 \rangle$	$\mathbf{T}_1 \leq \mathbf{T}_2$
$\langle \mathbf{T}_1 \rangle \leq \langle \mathcal{F}_1 \rangle$	$\mathbf{T}_1 \leq \mathcal{F}_1$
$\langle \mathbf{T}_1 \rangle$ is join prime	$\kappa(\mathbf{T}_1) = \{\langle \mathbf{T}_5 \rangle\}$
$\langle \mathbf{T}_2 \rangle \leq \langle \mathbf{R}_7 \rangle$	$\mathbf{T}_2 \leq \mathbf{R}_7$
$\langle \mathbf{T}_2 \rangle \leq \langle \mathbf{T}_1 \rangle \vee \langle \mathbf{T}_5 \rangle$	$\mathbf{T}_2 \leq \mathbf{T}_1 \times \mathbf{T}_5$
$\langle \mathbf{T}_3 \rangle \leq \langle \mathbf{R}_j \rangle$ for $j = 2, 4, 6$	$\mathbf{T}_3 \leq \mathbf{R}_j$ for $j = 2, 4, 6$
$\langle \mathbf{T}_3 \rangle$ is join prime	$\kappa(\mathbf{T}_3) = \{\langle gx \approx 0 \to fx \approx 0 \rangle\}$
$\langle \mathbf{T}_4 \rangle \leq \langle \mathbf{R}_j \rangle$ for $j = 3, 5, 6$	$\mathbf{T}_4 \leq \mathbf{R}_j$ for $j = 3, 5, 6$
$\langle \mathbf{T}_4 \rangle$ is join prime	$\kappa(\mathbf{T}_4) = \{\langle fx \approx gx \to fx \approx 0 \rangle\}$
$\langle \mathbf{T}_5 \rangle$ is join prime	$\kappa(\mathbf{T}_5) = \{\langle fx \approx gx \approx 0 \to x \approx 0 \rangle\}$
$\langle \mathcal{F}_1 \rangle \leq \langle \mathbf{R}_1 \rangle$	$\mathcal{F}_1 \leq \mathbf{R}_1$
$\langle \mathcal{F}_1 \rangle \leq \langle \mathbf{R}_8 \rangle$	$\mathcal{F}_1 \leq \mathbf{R}_8$
$\langle \mathcal{F}_1 \rangle$ is join prime	$\kappa(\mathcal{F}_1) = \{\langle fx \approx 0 \rangle\}$
$\langle \mathbf{R}_1 \rangle \leq \langle \mathbf{R}_2 \rangle$	$\mathbf{R}_1 \leq \mathbf{T}_1 \times \mathbf{R}_2 \leq \mathbf{R}_2^2$
$\langle \mathbf{R}_1 \rangle \leq \langle \mathbf{R}_3 \rangle$	$\mathbf{R}_1 \leq \mathbf{T}_1 \times \mathbf{R}_3 \leq \mathbf{R}_3^2$
$\langle \mathbf{R}_1 \rangle \leq \langle \mathbf{T}_2 \rangle \vee \langle \mathcal{F}_1 \rangle \leq \langle \mathbf{R}_7 \rangle$	$\mathbf{R}_1 \leq \mathbf{T}_2 \times \mathcal{F}_1 \leq \mathbf{R}_7^2$
$\langle \mathbf{R}_1 \rangle \leq \langle \mathbf{T}_5 \rangle \vee \langle \mathcal{F}_1 \rangle$	$\mathbf{R}_1 \leq \mathbf{T}_5 \times \mathcal{F}_1$
$\langle \mathbf{R}_2 \rangle \leq \langle \mathbf{T}_3 \rangle \vee \langle \mathbf{T}_5 \rangle$	$\mathbf{R}_2 \leq \mathbf{T}_3 \times \mathbf{T}_5$
$\langle \mathbf{R}_3 \rangle \leq \langle \mathbf{T}_4 \rangle \vee \langle \mathbf{T}_5 \rangle$	$\mathbf{R}_3 \leq \mathbf{T}_4 \times \mathbf{T}_5$
$\langle \mathbf{R}_4 \rangle \leq \langle \mathbf{T}_3 \rangle \vee \langle \mathbf{T}_5 \rangle$	$\mathbf{R}_4 \leq \mathbf{T}_3 \times \mathbf{T}_5$
$\langle \mathbf{R}_5 \rangle \leq \langle \mathbf{T}_4 \rangle \vee \langle \mathbf{T}_5 \rangle$	$\mathbf{R}_5 \leq \mathbf{T}_4 \times \mathbf{T}_5$
$\langle \mathbf{R}_6 \rangle$ is join prime	$\kappa(\mathbf{R}_6) = \{\langle fx \approx gx \approx fy \ \& \ gy \approx 0$ $\to fx \approx 0 \rangle\}$
$\langle \mathbf{R}_7 \rangle \leq \langle \mathbf{R}_4 \rangle$	$\mathbf{R}_7 \leq \mathbf{T}_1 \times \mathbf{R}_4 \leq \mathbf{R}_4^2$
$\langle \mathbf{R}_7 \rangle \leq \langle \mathbf{R}_5 \rangle$	$\mathbf{R}_7 \leq \mathbf{T}_1 \times \mathbf{R}_5 \leq \mathbf{R}_5^2$
$\langle \mathbf{R}_7 \rangle \leq \langle \mathbf{T}_5 \rangle \vee \langle \mathcal{F}_1 \rangle$	$\mathbf{R}_7 \leq \mathbf{T}_5 \times \mathcal{F}_1$
$\langle \mathbf{R}_8 \rangle \leq \langle \mathbf{T}_4 \rangle$	$\mathbf{R}_8 \leq \mathbf{T}_1 \times \mathbf{T}_4 \leq \mathbf{T}_4^2$
$\langle \mathbf{R}_8 \rangle \leq \langle \mathbf{P}_2 \rangle$	$\mathbf{R}_8 \leq \mathbf{T}_1 \times \mathbf{P}_2 \leq \mathbf{P}_2^2$
$\langle \mathbf{R}_8 \rangle$ is join prime	$\kappa(\mathbf{R}_8) = \{\langle fx \approx gy \to fy \approx 0 \rangle\}$
$\langle \mathbf{P}_2 \rangle \leq \langle \mathbf{T}_3 \rangle$	$\mathbf{P}_2 \leq \mathbf{T}_3^2$
$\langle \mathbf{P}_2 \rangle$ is join prime	$\kappa(\mathbf{P}_2) = \{\langle fx \approx gy \ \& \ gx \approx fy$ $\to fx \approx gx \rangle\}$

TABLE 5.3. Fun facts about small quasicriticals!

Table 5.4 lists the sets $\mathbf{O}(\mathbf{T}, \alpha)$ and $\mathbf{N}(\mathbf{T}, \alpha)$ for those pairs $(\mathbf{T}, \alpha)$ such that $\langle \varepsilon_{\mathbf{T}, \alpha} \rangle \in \kappa(\mathbf{T})$, i.e., $\langle \varepsilon_{\mathbf{T}, \alpha} \rangle = \widehat{\mathcal{U}}(\mathbf{T}, \alpha)$ is maximal in $\mathbf{L}_q(\mathcal{M})$ with respect to not containing $\langle \mathbf{T} \rangle$. Recall that $\langle \varepsilon_{\mathbf{T}, \alpha} \rangle \in \kappa(\mathbf{T})$ if the reflective closure $\gamma^{\mathbf{T}}(\alpha)$ is a minimal nonzero reflection congruence. Then $\mathbf{B}(\mathbf{T}, \alpha)$ consists of the factor algebras $\mathbf{T}/\varphi$ with $\varphi \not\geq \alpha$ in $\mathrm{Con}_{\mathcal{M}} \mathbf{T}$, $\mathbf{O}(\mathbf{T}, \alpha)$ is the collection of minimal members of $\mathbf{B}(\mathbf{T}, \alpha)$ with respect to subalgebra inclusion $\leq$, and $\mathbf{N}(\mathbf{T}, \alpha)$ is the set of minimal members of $\mathbf{B}(\mathbf{T}, \alpha)$ with respect to quasivariety

Alg.	Atom	$\langle \varepsilon_{\mathbf{T},\alpha} \rangle$	$O(\mathbf{T}, \alpha)$	$N(\mathbf{T}, \alpha)$
$\mathbf{T}_1$	$gx \equiv 0$	$\langle gx \approx 0 \rangle = \langle \mathbf{T}_5 \rangle$	$\mathbf{T}_1$	$\mathbf{T}_1$
$\mathbf{T}_2$	$gx \equiv 0$	$\langle gx \approx 0 \rangle = \langle \mathbf{T}_5 \rangle$	$\mathbf{T}_1$	$\mathbf{T}_1$
$\mathbf{T}_2$	$gx \equiv x$	$\langle fx \approx 0 \rightarrow gx \approx x \rangle$	$\mathbf{T}_2, \mathbf{T}_5$	$\mathbf{T}_2, \mathbf{T}_5$
$\mathbf{T}_3$	$fx \equiv 0$	$\langle gx \approx 0 \rightarrow fx \approx 0 \rangle$	$\mathbf{T}_3$	$\mathbf{T}_3$
$\mathbf{T}_4$	$fx \equiv 0$	$\langle fx \approx gx \rightarrow fx \approx 0 \rangle$	$\mathbf{T}_4$	$\mathbf{T}_4$
$\mathbf{T}_5$	$x \equiv 0$	$\langle fx \approx gx \approx 0 \rightarrow x \approx 0 \rangle$	$\mathbf{T}_5$	$\mathbf{T}_5$
$\mathcal{F}_1$	$fx \equiv 0$	$\langle fx \approx 0 \rangle = \langle \mathbf{T}_1 \rangle \vee \langle \mathbf{T}_5 \rangle$	$\mathbf{T}_3, \mathbf{T}_4, \mathcal{F}_1$	$\mathcal{F}_1$
$\mathbf{R}_1$	$fx \equiv 0$	$\langle fx \approx 0 \rangle = \langle \mathbf{T}_1 \rangle \vee \langle \mathbf{T}_5 \rangle$	$\mathbf{T}_3, \mathbf{T}_4, \mathcal{F}_1$	$\mathcal{F}_1$
$\mathbf{R}_1$	$x \equiv y$	$\langle fx \approx fy \ \& \ gx \approx gy \rightarrow x \approx y \rangle$	$\mathbf{T}_2, \mathbf{T}_5, \mathbf{R}_1,$ $\mathbf{R}_2, \mathbf{R}_3$	$\mathbf{T}_2, \mathbf{T}_5, \mathbf{R}_1$
$\mathbf{R}_2$	$fx \equiv 0$	$\langle gx \approx 0 \rightarrow fx \approx 0 \rangle$	$\mathbf{T}_3$	$\mathbf{T}_3$
$\mathbf{R}_2$	$x \equiv y$	$\langle fx \approx fy \ \& \ gx \approx gy \approx 0$ $\rightarrow x \approx y \rangle$	$\mathbf{T}_5, \mathbf{R}_2$	$\mathbf{T}_5, \mathbf{R}_2$
$\mathbf{R}_3$	$fx \equiv 0$	$\langle fx \approx gx \rightarrow fx \approx 0 \rangle$	$\mathbf{T}_4$	$\mathbf{T}_4$
$\mathbf{R}_3$	$x \equiv y$	$\langle fx \approx fy \approx gx \approx gy \rightarrow x \approx y \rangle$	$\mathbf{T}_5, \mathbf{R}_3$	$\mathbf{T}_5, \mathbf{R}_3$
$\mathbf{R}_4$	$gx \equiv 0$	$\langle gx \approx 0 \rightarrow fx \approx 0 \rangle$	$\mathbf{T}_3$	$\mathbf{T}_3$
$\mathbf{R}_4$	$gx \equiv x$	$\langle fx \approx gy \approx 0 \ \& \ gx \approx fy$ $\rightarrow gx \approx x \rangle$	$\mathbf{T}_5, \mathbf{R}_4$	$\mathbf{T}_5, \mathbf{R}_4$
$\mathbf{R}_5$	$gy \equiv 0$	$\langle fx \approx gx \rightarrow fx \approx 0 \rangle$	$\mathbf{T}_4$	$\mathbf{T}_4$
$\mathbf{R}_5$	$gy \equiv y$	$\langle fx \approx gx \approx gy \ \& \ fy \approx 0$ $\rightarrow gy \approx y \rangle$	$\mathbf{T}_5, \mathbf{R}_5$	$\mathbf{T}_5, \mathbf{R}_5$
$\mathbf{R}_6$	$fx \equiv 0$	$\langle fx \approx gx \approx fy \ \& \ gy \approx 0$ $\rightarrow fx \approx 0 \rangle$	$\mathbf{R}_6$	$\mathbf{R}_6$
$\mathbf{R}_7$	$fx \equiv 0$	$\langle fx \approx 0 \rangle = \langle \mathbf{T}_1 \rangle \vee \langle \mathbf{T}_5 \rangle$	$\mathbf{T}_3, \mathbf{T}_4, \mathcal{F}_1$	$\mathcal{F}_1$
$\mathbf{R}_7$	$fx \equiv y$	$\langle fx \approx gy \ \& \ fy \approx 0 \rightarrow fx \approx y \rangle$	$\mathbf{T}_5, \mathbf{R}_4, \mathbf{R}_5, \mathbf{R}_7$	$\mathbf{T}_5, \mathbf{R}_7$
$\mathbf{R}_8$	$fy \equiv 0$	$\langle fx \approx gy \rightarrow fy \approx 0 \rangle$	$\mathbf{T}_3, \mathbf{T}_4, \mathbf{R}_8, \mathbf{P}_2$	$\mathbf{R}_8$
$\mathbf{P}_2$	$fx \equiv gx$	$\langle fx \approx gy \ \& \ gx \approx fy$ $\rightarrow fx \approx gx \rangle$	$\mathbf{T}_3, \mathbf{P}_2$	$\mathbf{P}_2$

TABLE 5.4. Sins of omission. Recall that an algebra $\mathbf{A} \in \mathcal{M}$ satisfies $\varepsilon_{\mathbf{T},\alpha}$ if and only if $\mathbf{S} \not\leq \mathbf{A}$ for all $S \in O(\mathbf{T}, \alpha)$, and a quasivariety $\mathcal{Q} \leq \mathcal{M}$ satisfies $\varepsilon_{\mathbf{T},\alpha}$ if and only if $\mathbf{U} \notin \mathcal{Q}$ for all $U \in N(\mathbf{T}, \alpha)$ (Theorem 3.4). In cases where $\langle \varepsilon_{\mathbf{T},\alpha} \rangle = \langle \varepsilon_{\mathbf{S},\beta} \rangle$, the simpler of the two equivalent quasi-equations is given in the third column.

inclusion $\leq_q$. Theorem 3.4 then states that a finite algebra $\mathbf{A} \in \mathcal{M}$ satisfies $\varepsilon_{\mathbf{T},\alpha}$ if and only if $\mathbf{U} \not\leq \mathbf{A}$ for every $\mathbf{U}$ in $\mathbf{O}(\mathbf{T}, \alpha)$, or equivalently, $\mathbf{R} \notin \langle \mathbf{A} \rangle$ for every $\mathbf{R} \in \mathbf{N}(\mathbf{T}, \alpha)$.

In (3) and (4) we are looking for algebras $\mathbf{S}$ such that $\mathbf{S} \notin \langle \mathbf{T} \rangle$, so that $\langle \mathbf{T} \rangle$ satisfies $\varepsilon_{\mathbf{S},\beta}$ for some atom β of Con $\mathbf{S}$. Part (3) asks for minimality with respect to quasivariety inclusion order, while (4) asks for minimality with respect to subalgebra inclusion. With the order-correct partial diagram of $L_q(\mathcal{M})$ in Figure 5.5, we can identify small quasicritical algebras with $\langle \mathbf{S} \rangle \not\leq \langle \mathbf{T} \rangle$ for part (3), that are at least minimal among the quasivarieties generated by small quasicriticals, if not overall. Likewise, using Figure 5.3, we can find the small quasicritical algebras that are minimal with respect to $\mathbf{S} \not\leq \mathbf{T}$; those that also satisfy $\mathbf{S} \notin \langle \mathbf{T} \rangle$ are listed in part (4).

By Theorem 3.10, if we can find a complete list for (3) or (4), then the corresponding quasi-equations $\varepsilon_{\mathbf{S},\beta}$ satisfied by $\mathbf{T}$ form a basis for $\langle \mathbf{T} \rangle$. For some small quasicritical algebras in $\mathcal{M}$, a basis for $\langle \mathbf{T} \rangle$ can be found this way. For others, we may know that $\langle \mathbf{T} \rangle$ is finitely based without knowing whether the lists in (3) and (4) are complete. Regardless of whether $\langle \mathbf{T} \rangle$ is finitely based, the information in (3) allows us to address (1), the quasi-equations satisfied by $\mathbf{T}$.

This completes our outline of the calculations.

A *unary algebra with* 0 is an algebra with unary operations $f_1, \ldots, f_k$ and a constant 0 satisfying $f_j(0) \approx 0$ for all j. Such an algebra $\mathbf{A}$ has 2-*element range* if there exists an element $1 \in A$ such that $f_j(x) \in \{0, 1\}$ for all j and all $x \in A$. Let $\overline{f}(x)$ denote the vector $(f_1(x), \ldots, f_k(x))$. Then Rows$(\mathbf{A})$ is the collection of vectors $\{\overline{f}(x) : x \in A\}$. This just consists of the set of rows of the operation table of $\mathbf{A}$. Note that for a unary algebra with 0 and 2-element range, Rows$(\mathbf{A}) \subseteq \mathbf{2}^k$.

Now unary algebras with 0 and 2-element range do not form a quasivariety, but one can talk about those contained in a given quasivariety. Given a unary algebra $\mathbf{A}$, we say that Rows$(\mathbf{A})$ is *uniquely witnessed* if $\mathbf{A}$ satisfies $\overline{f}(x) \approx \overline{f}(y) \rightarrow x \approx y$.

The main result of Casperson *et al.* [43] can be stated as follows.

THEOREM 5.2. *Let* $\mathbf{A}$ *be a unary algebra with* 0 *and 2-element range.*

(1) *The quasivariety* $\langle \mathbf{A} \rangle$ *is finitely based if and only if* Rows$(\mathbf{A})$ *is an order ideal of* $\mathbf{2}^k$.

(2) *If* Rows$(\mathbf{A})$ *is an order ideal of* $\mathbf{2}^k$ *and uniquely witnessed, then* $\langle \mathbf{A} \rangle$ *has a basis of 2-variable quasi-identities.*

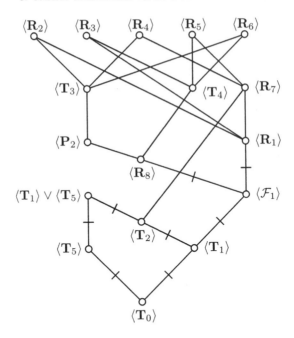

FIGURE 5.5. The bottom of $L_q(\mathcal{M})$, derived from the information in Figure 5.3 and Table 5.3. The diagram gives the order, but not joins and meets. Ticks indicate covers.

Looking at the small quasicritical algebras in our variety $\mathcal{M}$, we see that $\mathcal{F}_1$, $\mathbf{R}_1$, $\mathbf{R}_7$, $\mathbf{R}_8$, and $\mathbf{P}_2$ do not have a 2-element range, but the remaining ones do. The rows form an order ideal for some of the small quasicriticals in $\mathcal{M}$ with a 2-element range: $\mathbf{T}_1$, $\mathbf{T}_2$, $\mathbf{T}_3$, $\mathbf{T}_5$, $\mathbf{R}_2$, $\mathbf{R}_4$, $\mathbf{R}_6$. So for these algebras the corresponding quasivarieties $\langle \mathbf{A} \rangle$ are finitely based. On the other hand, the rows do not form an order ideal for $\mathbf{T}_4$, $\mathbf{R}_3$, and $\mathbf{R}_5$, and the quasivarieties they generate are not finitely based.

Of the finitely based quasivarieties, the rows are uniquely witnessed for $\mathbf{T}_1$, $\mathbf{T}_3$, and $\mathbf{R}_6$. Their quasivarieties have a basis of 2-variable quasi-identities, which we can find using Theorem 3.10, because the minimal algebras not in $\langle \mathbf{A} \rangle$ are 2-generated. Indeed, we will see in Proposition 5.13(12) that for $\mathbf{S} \in \mathcal{M}$, Rows($\mathbf{S}$) is uniquely witnessed if and only if none of $\mathbf{T}_2$, $\mathbf{T}_5$, and $\mathbf{R}_1$ is a subalgebra of $\mathbf{S}$. It turns out that some of the other small quasivarieties have a 2-variable basis as well.

Let us note a decomposition that plays a crucial role for unary algebras with 0. The *glued sum* of two algebras $\mathbf{A}$, $\mathbf{B} \in \mathcal{M}$, denoted $\mathbf{A} * \mathbf{B}$, is the algebra whose universe is $A \cup B$ with $A \cap B = \{0\}$, with its operations inherited from $\mathbf{A}$ and $\mathbf{B}$. Note that $\mathbf{A}, \mathbf{B} \leq \mathbf{A} * \mathbf{B}$ and that it is a subdirect product of $\mathbf{A}$ and $\mathbf{B}$. Thus $\mathbf{A} * \mathbf{B}$ is certainly not quasicritical when the factors are nontrivial.

As an example, for $k > 1$ the free algebra $\mathcal{F}_k \cong \mathcal{F}_1 * \cdots * \mathcal{F}_1$ with k factors. Thus $\mathcal{F}_k \in \langle \mathcal{F}_1 \rangle$.

The next lemma also uses this construction.

LEMMA 5.3. *The only quasicritical algebras of* $\mathcal{M}$ *in the quasivariety* $\langle fx \approx 0 \rangle$ *are* $\mathbf{T}_1$, $\mathbf{T}_2$, *and* $\mathbf{T}_5$. *Moreover,* $\langle fx \approx 0 \rangle = \langle \mathbf{T}_1 \rangle \vee \langle \mathbf{T}_5 \rangle$.

PROOF. Let $\mathbf{A}$ be any finite algebra in $\langle fx \approx 0 \rangle$. Then the structure of $\mathbf{A}$ is determined by the action of the operation g, which satisfies $g^2 x \approx gx$. Thus it is easy to see that

(i) $\mathbf{A}$ is generated by $G = \{a \in \mathbf{A} : a \neq g(b) \text{ for all } b \neq a\}$;
(ii) $\mathbf{A}$ is the set union of the subalgebras $\mathrm{Sg}(a)$ for $a \in G$;
(iii) each of these is isomorphic to $\mathbf{T}_1$, $\mathbf{T}_2$, or $\mathbf{T}_5$ (according as to whether $g(a) = a$, $a \neq g(a) \neq 0$ or $g(a) = 0$);
(iv) pairwise, for $a \neq b$, $\mathrm{Sg}(a) \cap \mathrm{Sg}(b)$ is either $\{0\}$ or $\{0, g(a)\}$, the latter case occurring when $g(a) = g(b)$.

It follows that $\mathbf{A} = \mathbf{S}_1 * \cdots * \mathbf{S}_m$ where each subalgebra $\mathbf{S}_i$ is isomorphic to $\mathbf{T}_1$, $\mathbf{T}_2$, $\mathbf{T}_5$ or a "fan" $\{a_1, \ldots, a_k, c, 0\}$ with $k \geq 2$ and $c = f(a_j)$ for all j. These fans can be embedded in $\mathbf{T}_2^k$ and so are not quasicritical: in the congruence lattice of a fan, $\Delta = \bigcap_{1 \leq i \leq k} \psi_i$ where ψ_i is the congruence whose only non-singleton block is $\{a_j : j \neq i\} \cup \{c\}$.

Thus $\langle fx \approx 0 \rangle$ contains only the quasicriticals $\mathbf{T}_1$, $\mathbf{T}_2$, $\mathbf{T}_5$. Moreover, since $\mathbf{T}_2 \leq \mathbf{T}_1 \times \mathbf{T}_5$, we have $\langle fx \approx 0 \rangle = \langle \mathbf{T}_1 \rangle \vee \langle \mathbf{T}_5 \rangle$.   □

LEMMA 5.4. *If a quasivariety* $\mathfrak{Q} \leq \mathcal{M}$ *does not satisfy* $fx \approx 0$, *then* $\mathcal{F}_1 \in \mathfrak{Q}$.

PROOF. Such a quasivariety $\mathfrak{Q}$ must contain a 1-generated quasicritical algebra not isomorphic to $\mathbf{T}_1$, $\mathbf{T}_2$, or $\mathbf{T}_5$. By looking at the embeddings of Figure 5.3, we see that $\mathfrak{Q}$ must contain $\mathbf{T}_1$ and at least one of $\mathbf{T}_3$, $\mathbf{T}_4$, and $\mathcal{F}_1$. From the congruence lattice of $\mathcal{F}_1$, given in Figure 5.4, we see that $\mathcal{F}_1 \leq \mathbf{T}_1 \times \mathbf{T}_3$ and $\mathcal{F}_1 \leq \mathbf{T}_1 \times \mathbf{T}_4$. Thus $\mathcal{F}_1 \in \mathfrak{Q}$.   □

The quasivariety $\langle \mathbf{T}_5 \rangle$ warrants special attention.

LEMMA 5.5. $\langle \mathbf{T}_5 \rangle = \langle gx \approx 0 \rangle = \langle fx \approx gx \rangle < \langle fx \approx 0 \rangle$.

PROOF. If $gx \approx 0$ holds in a quasivariety $\mathfrak{Q} \leq \mathcal{M}$, then using the substitution $x \mapsto fx$ and the equations of $\mathcal{M}$ yields $fx \approx g(fx) \approx 0$, and in particular $fx \approx gx$. If $fx \approx gx$ holds in $\mathfrak{Q}$, then the substitution $x \mapsto gx$ gives $gx \approx g(gx) \approx g(fx) \approx 0$. So the equations are equivalent in $\mathcal{M}$ and imply $fx \approx 0$. In view of Lemma 5.3, $\mathbf{T}_5$ is the only quasicritical algebra satisfying those equations.   □

Let us begin our case-by-case analysis of the quasivarieties generated by the small quasicritical algebras in $\mathcal{M}$. We start with $\mathbf{T}_1$, which is fairly straightforward because it is a simple algebra that satisfies $fx \approx 0$.

PROPOSITION 5.6. *The algebra* $\mathbf{T}_1$ *satisfies the following.*

(1) $\langle \mathbf{T}_1 \rangle = \langle gx \approx x \rangle$.
(2) *The only quasicritical algebra in* $\langle \mathbf{T}_1 \rangle$ *is* $\mathbf{T}_1$.
(3) $\mathfrak{Q} \leq \langle \mathbf{T}_1 \rangle$ *if and only if* $\mathfrak{Q}$ *omits* $\mathbf{T}_2$, $\mathbf{T}_5$, *and* $\mathcal{F}_1$.
(4) $\mathbf{S} \in \langle \mathbf{T}_1 \rangle$ *if and only if* $\mathbf{S}$ *omits* $\mathbf{T}_2$, $\mathbf{T}_3$, $\mathbf{T}_4$, $\mathbf{T}_5$, *and* $\mathcal{F}_1$.
(5) $\langle \mathbf{T}_1 \rangle_* = \langle \mathbf{T}_0 \rangle$.
(6) $\mathbf{T}_1$ *is simple, so its reflection congruences are* Δ *and* ∇.
(7) $\gamma^{\mathbf{T}_1}(\nabla) = \nabla$ *is the congruence* $x \equiv 0$.
(8) $\operatorname{diag}(\mathbf{T}_1)$ *is* $fx \approx 0$ & $gx \approx x$.
(9) $\mathcal{E}(\mathbf{T}_1) = \{\varepsilon_{\mathbf{T}_1, x \equiv 0}\}$.
(10) $\kappa(\mathbf{T}_1) = \lambda(\mathbf{T}_1) = \{\widehat{\mathcal{U}}(\mathbf{T}_1, x \equiv 0)\}$.
(11) $\widehat{\mathcal{U}}(\mathbf{T}_1, x \equiv 0) = \langle gx \approx 0 \rangle = \langle \mathbf{T}_5 \rangle$.
(12) $\mathbf{S} \in \widehat{\mathcal{U}}(\mathbf{T}_1, x \equiv 0)$ *if and only if* $\mathbf{S}$ *omits* $\mathbf{T}_1$.
(13) *See* Figure 5.9 *for the segment of* $L_q(\mathcal{M})$ *related to* $\mathbf{T}_1$.

PROOF. The algebra $\mathbf{T}_1$ satisfies the equation $gx \approx x$. Applying f and using one of the defining equations of $\mathcal{M}$ gives $0 \approx fgx \approx fx$, so $gx \approx x$ implies $fx \approx 0$. In view of Lemma 5.3, $\mathbf{T}_1$ is the only quasicritical algebra that satisfies these two equations. Furthermore $\langle \mathbf{T}_1 \rangle_* = \langle \mathbf{T}_0 \rangle$.

The congruence lattice of $\mathbf{T}_1$ is a 2-element chain with the non-identity congruence defined by $x \equiv 0$. Thus $\kappa(\mathbf{T}_1) = \lambda(\mathbf{T}_1) = \{\widehat{\mathcal{U}}(\mathbf{T}_1, x \equiv 0)\}$, and Lemma 3.6 gives (12).

As the defining relations for $\mathbf{T}_1$ from $\mathcal{F}_1$ are $fx \equiv 0$ and $gx \equiv x$, the quasi-equation $\varepsilon_{\mathbf{T}_1, x \equiv 0}$ is $gx \approx x$ & $fx \approx 0 \to x \approx 0$. The substitution $x \mapsto gx$ applied to $\varepsilon_{\mathbf{T}_1, x \equiv 0}$ yields $g^2 x \approx gx$ & $fgx \approx 0 \to gx \approx 0$. Since the hypothesis of this is always true in $\mathcal{M}$, the equation $gx \approx 0$ holds in $\widehat{\mathcal{U}}(\mathbf{T}_1, x \equiv 0)$. Conversely, assume that an algebra $\mathbf{A}$ satisfies $gx \approx 0$ and that the hypothesis of $\varepsilon_{\mathbf{T}_1, x \equiv 0}$ for the element $a \in A$. This means that $g(a) = 0$ and $g(a) = a$, so $a = 0$, which is the conclusion of $\varepsilon_{\mathbf{T}_1, x \equiv 0}$. Thus $gx \approx 0$ and $\varepsilon_{\mathbf{T}_1, x \equiv 0}$ are equivalent in $\mathcal{M}$, which is the first part of (11). Lemma 5.5 then gives $\langle gx \approx 0 \rangle = \langle \mathbf{T}_5 \rangle$. □

PROPOSITION 5.7. *The algebra* $\mathbf{T}_2$ *satisfies the following:*

(1) $\langle \mathbf{T}_2 \rangle = \langle fx \approx 0, \ gx \approx 0 \to x \approx 0 \rangle$.
(2) *The quasicritical algebras in* $\langle \mathbf{T}_2 \rangle$ *are* $\mathbf{T}_1$ *and* $\mathbf{T}_2$.
(3) $\mathfrak{Q} \leq \langle \mathbf{T}_2 \rangle$ *if and only if* $\mathfrak{Q}$ *omits* $\mathbf{T}_5$ *and* $\mathcal{F}_1$.
(4) $\mathbf{S} \in \langle \mathbf{T}_2 \rangle$ *if and only if* $\mathbf{S}$ *omits* $\mathbf{T}_3$, $\mathbf{T}_4$, $\mathbf{T}_5$, *and* $\mathcal{F}_1$.
(5) $\langle \mathbf{T}_2 \rangle_* = \langle \mathbf{T}_1 \rangle$.
(6) *All four congruences of* $\mathbf{T}_2$ *are reflection congruences:* Δ, ∇, $gx \equiv x$, $gx \equiv 0$.
(7) $\gamma^{\mathbf{T}_2}(\alpha) = \alpha$ *for the atoms* $gx \equiv x$ *and* $gx \equiv 0$.
(8) $\operatorname{diag}(\mathbf{T}_2)$ *is* $fx \approx 0$.
(9) $\mathcal{E}(\mathbf{T}_2) = \{\varepsilon_{\mathbf{T}_2, gx \equiv x}, \varepsilon_{\mathbf{T}_2, gx \equiv 0}\}$.
(10) $\kappa(\mathbf{T}_2) = \lambda(\mathbf{T}_2) = \{\widehat{\mathcal{U}}(\mathbf{T}_2, gx \equiv x), \widehat{\mathcal{U}}(\mathbf{T}_2, gx \equiv 0)\}$.
(11) $\widehat{\mathcal{U}}(\mathbf{T}_2, gx \equiv x)$ *is above* $\langle \mathbf{R}_2 \rangle \vee \langle \mathbf{R}_3 \rangle \vee \langle \mathbf{R}_6 \rangle$ *but not above* $\langle \mathbf{T}_5 \rangle$,
      $\widehat{\mathcal{U}}(\mathbf{T}_2, gx \equiv 0) = \langle gx \approx 0 \rangle = \langle \mathbf{T}_5 \rangle$.

(12) $\mathbf{S} \in \widehat{\mathcal{U}}(\mathbf{T}_2, gx \equiv x)$ *if and only if* $\mathbf{S}$ *omits* $\mathbf{T}_2$ *and* $\mathbf{T}_5$,
    $\mathbf{S} \in \widehat{\mathcal{U}}(\mathbf{T}_2, gx \equiv 0)$ *if and only if* $\mathbf{S}$ *omits* $\mathbf{T}_1$.
(13) *See* Figure 5.9 *for the segment of* $\mathrm{L}_q(\mathcal{M})$ *related to* $\mathbf{T}_2$.

PROOF. The algebra $\mathbf{T}_2$ satisfies $fx \approx 0$ and $gx \approx 0 \to x \approx 0$, so we can use Lemma 5.3 to see that the only quasicritical algebras in $\langle \mathbf{T}_2 \rangle$ are $\mathbf{T}_1$ and $\mathbf{T}_2$.

The congruence lattice of $\mathbf{T}_2$ is isomorphic to $\mathbf{2} \times \mathbf{2}$ with atoms given by $gx \equiv 0$ and $gx \equiv x$; it appears as an upper quotient in Figure 5.4. The quasi-equation $\varepsilon_{\mathbf{T}_2, gx \equiv 0}$, which is $fx \approx 0 \to gx \approx 0$, is equivalent to $gx \approx 0$. To see this apply the substitution $x \mapsto gx$ for one direction, and note that an equation always implies a quasi-equation whose conclusion is that equation. Thus $\widehat{\mathcal{U}}(\mathbf{T}_2, gx \equiv 0) = \langle gx \approx 0 \rangle$, which by Lemma 5.5 is $\langle \mathbf{T}_5 \rangle$.

The largest congruence in Con $\mathbf{T}_2$ that is not above $gx \equiv x$ is $gx \equiv 0$. By Theorem 3.4, $\mathbf{S} \in \langle \varepsilon_{\mathbf{T}_2, gx \equiv x} \rangle = \widehat{\mathcal{U}}(\mathbf{T}_2, gx \equiv x)$ if and only if $\mathbf{S}$ omits $\mathbf{T}_2/\Delta$ and $\mathbf{T}_2/(gx \equiv 0)$, that is, if and only if $\mathbf{S}$ omits $\mathbf{T}_2$ and $\mathbf{T}_5$. Similarly, as $\mathbf{T}_1$ is a subalgebra of $\mathbf{T}_2$, we have $\mathbf{S} \in \langle \varepsilon_{\mathbf{T}_2, gx \equiv 0} \rangle = \widehat{\mathcal{U}}(\mathbf{T}_2, gx \equiv 0)$ if and only if $\mathbf{S}$ omits $\mathbf{T}_1$.   □

Let us temporarily postpone the quasivariety $\langle \mathbf{T}_3 \rangle$ to Section 5.4. Note that $\mathbf{T}_3 = \mathbf{M}$, which is the algebra generating the variety $\mathcal{M}$. The quasivariety $\langle \mathbf{T}_3 \rangle$ is exceedingly interesting because it contains all the quasicritical algebras $\mathbf{P}_n$ that are used to show that $\mathcal{M}$ contains uncountably many subquasivarieties.

PROPOSITION 5.8. *The algebra* $\mathbf{T}_4$ *satisfies the following:*

(1) $\langle \mathbf{T}_4 \rangle$ *is not finitely based. Its quasi-identities include*

$$(\varepsilon_{\mathbf{R}_1, x \equiv y}) \qquad fx \approx fy \ \& \ gx \approx gy \to x \approx y,$$
$$(\varepsilon_{\mathbf{P}_2, fx \equiv 0}) \qquad fx \approx gy \ \& \ gx \approx fy \to fx \approx 0,$$

and the quasi-equations $\varepsilon_{\mathbf{P}_n, fx_1 \equiv gx_1}$ *for* $n \geq 2$ *of Theorem 5.26.*
(2) *The quasicritical algebras in* $\langle \mathbf{T}_4 \rangle$ *include* $\mathbf{T}_1$, $\mathcal{F}_1$, *and* $\mathbf{R}_8$.
(3) *If* $\mathcal{Q} \leq \langle \mathbf{T}_4 \rangle$, *then* $\mathcal{Q}$ *omits* $\mathbf{T}_2$, $\mathbf{T}_5$, $\mathbf{R}_1$, *and* $\mathbf{P}_2$.
(4) *If* $\mathbf{S} \in \langle \mathbf{T}_4 \rangle$, *then* $\mathbf{S}$ *omits* $\mathbf{T}_2$, $\mathbf{T}_3$, $\mathbf{T}_5$, $\mathbf{R}_1$, $\mathbf{R}_3$, *and* $\mathbf{P}_2$.
(5) $\langle \mathbf{T}_4 \rangle_* = \langle \mathbf{T}_4 \rangle \wedge \widehat{\mathcal{U}}(\mathbf{T}_4, fx \equiv 0) \geq \langle \mathbf{R}_8 \rangle$.
(6) *All three congruences of* $\mathbf{T}_4$ *are reflection congruences:* Δ, ∇, $fx \equiv 0$.
(7) $\gamma^{\mathbf{T}_4}(fx \equiv 0) = \rho^{\mathbf{T}_4}_{\langle fx \approx 0 \rangle}$ *is* $fx \equiv 0$.
(8) $\mathrm{diag}(\mathbf{T}_4)$ *is* $fx \approx gx$.
(9) $\mathcal{E}(\mathbf{T}_4) = \{\varepsilon_{\mathbf{T}_4, fx \equiv 0}\}$.
(10) $\kappa(\mathbf{T}_4) = \lambda(\mathbf{T}_4) = \{\widehat{\mathcal{U}}(\mathbf{T}_4, fx \equiv 0)\}$.
(11) $\widehat{\mathcal{U}}(\mathbf{T}_4, fx \equiv 0) = \langle fx \approx gx \to fx \approx 0 \rangle$.
(12) $\mathbf{S} \in \widehat{\mathcal{U}}(\mathbf{T}_4, fx \equiv 0)$ *if and only if* $\mathbf{S}$ *omits* $\mathbf{T}_4$.

PROOF. The congruence lattice of $\mathbf{T}_4$ is the 3-element chain which appears as an upper quotient in Figure 5.4. Its atom is determined by $fx \equiv 0$, and $\mathbf{T}_4/(fx \equiv 0)$ is isomorphic to $\mathbf{T}_5$.

As observed earlier, $\langle \mathbf{T}_4 \rangle$ is not finitely based by Theorem 5.2.   □

There are quasivarieties between $\langle \mathbf{R}_8 \rangle$ and $\langle \mathbf{T}_4 \rangle$ in $L_q(\mathcal{M})$. For example, let $\mathbf{P}_3^\dagger$ be the algebra $\mathcal{F}_3/\psi$ where

$$\psi = (fx_2 \equiv fx_1) \vee (fx_3 \equiv gx_2) \vee (gx_1 \equiv gx_3),$$

which is the first algebra drawn in Figure 5.6. Then the Universal Algebra Calculator shows that $\mathbf{P}_3^\dagger$ is quasicritical, while one can easily check that $\mathbf{R}_8 \leq \mathbf{P}_3^\dagger \leq \mathbf{T}_1 \times \mathbf{T}_4^2$. Hence $\langle \mathbf{R}_8 \rangle < \langle \mathbf{P}_3^\dagger \rangle < \langle \mathbf{T}_4 \rangle$. On the other hand, $\mathbf{P}_3^\dagger \notin \langle \mathbf{T}_3 \rangle$.

An odd thing happens when we extend this construction. For $k \geq 4$, define $\mathbf{P}_k^\dagger$ to be the algebra $\mathcal{F}_k/\psi$ where

$$\psi = (fx_2 \equiv fx_1) \vee (fx_3 \equiv gx_2) \vee \cdots \vee (fx_k \equiv gx_{k-1}) \vee (gx_1 \equiv gx_k).$$

The algebras $\mathbf{P}_4^\dagger$ and $\mathbf{P}_5^\dagger$ are drawn in Figure 5.6. But in contrast to $\mathbf{P}_3^\dagger$, for $k \geq 4$ these algebras satisfy $\langle \mathbf{R}_8 \rangle < \langle \mathbf{P}_k^\dagger \rangle < \langle \mathbf{T}_3 \rangle$, while $\mathbf{P}_k^\dagger \notin \langle \mathbf{T}_4 \rangle$.

PROPOSITION 5.9. *The algebra $\mathbf{T}_5$ satisfies the following:*

(1) $\langle \mathbf{T}_5 \rangle = \langle gx \approx 0 \rangle = \langle fx \approx gx \rangle$.

(2) $\mathbf{T}_5$ *is the only quasicritical algebra in* $\langle \mathbf{T}_5 \rangle$.

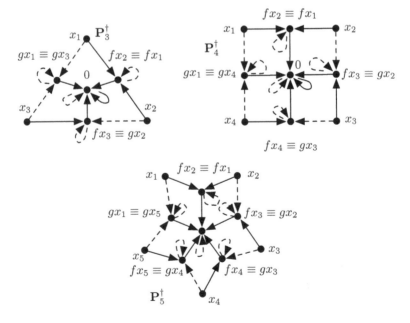

FIGURE 5.6. The algebras $\mathbf{P}_3^\dagger$, $\mathbf{P}_4^\dagger$, and $\mathbf{P}_5^\dagger$

(3) $\mathcal{Q} \leq \langle \mathbf{T}_5 \rangle$ if and only if $\mathcal{Q}$ omits $\mathbf{T}_1$.

(4) $\mathbf{S} \in \langle \mathbf{T}_5 \rangle$ if and only if $\mathbf{S}$ omits $\mathbf{T}_1$.

(5) $\langle \mathbf{T}_5 \rangle_* = \langle \mathbf{T}_0 \rangle$.

(6) $\mathbf{T}_5$ is simple, so its reflection congruences are Δ and ∇.

(7) $\gamma^{\mathbf{T}_5}(\nabla) = \nabla$ is $x \equiv 0$.

(8) $\mathrm{diag}(\mathbf{T}_5)$ is $fx \approx gx \approx 0$.

(9) $\mathcal{E}(\mathbf{T}_5) = \{\varepsilon_{\mathbf{T}_5, x \equiv 0}\}$.

(10) $\kappa(\mathbf{T}_5) = \lambda(\mathbf{T}_5) = \{\widehat{\mathcal{U}}(\mathbf{T}_5, x \equiv 0)\}$.

(11) $\widehat{\mathcal{U}}(\mathbf{T}_5, x \equiv 0) = \langle fx \approx gx \approx 0 \to x \approx 0 \rangle$.

(12) $\mathbf{S} \in \widehat{\mathcal{U}}(\mathbf{T}_5, x \equiv 0)$ if and only if $\mathbf{S}$ omits $\mathbf{T}_5$.

(13) See Figure 5.9 for the segment of $\mathrm{L}_q(\mathcal{M})$ related to $\mathbf{T}_5$.

PROOF. $\mathbf{T}_5$ is a 2-element simple algebra, with its top congruence ∇ defined by $x \equiv 0$. Lemmas 5.3 and 5.5 make the claims straightforward. Note that $\widehat{\mathcal{U}}(\mathbf{T}_5, x \equiv 0)$ contains all the small quasicritical algebras of $\mathcal{M}$ except $\mathbf{T}_5$. □

The algebra $\mathbf{T}_5$ has another nice property: it is not a subalgebra of any other quasicritical algebra in $\mathcal{M}$ (cf. Figure 5.3).

LEMMA 5.10. If $\mathbf{T}_5 \leq \mathbf{S}$, then $\mathbf{S} = \mathbf{T}_5 * \mathbf{B}$ for some $\mathbf{B} \leq \mathbf{S}$.

PROOF. Let $\mathbf{S} \in \mathcal{M}$ and $a \in S$ with $\mathrm{Sg}(a) \cong \mathbf{T}_5$. Then $f(a) = g(a) = 0$. For any $b \in S \setminus \{a\}$ we can have neither $f(b) = a$ nor $g(b) = a$, because $\mathcal{M}$ satisfies $gf(x) \approx f(x)$ and $g^2(x) \approx g(x)$. Hence $\mathbf{S} = \{a, 0\} * \mathbf{B}$ for $\mathbf{B} = \mathbf{S} \setminus \{a\}$. □

It turns out that $\langle \mathcal{F}_1 \rangle$ is also very small and has a 2-variable basis.

LEMMA 5.11. $\langle \mathcal{F}_1 \rangle$ has the quasi-equational basis

$$(\varepsilon_{\mathbf{R}_1, x \equiv y}) \qquad fx \approx fy \ \& \ gx \approx gy \to x \approx y,$$

$$(\varepsilon_{\mathbf{R}_8, fy \equiv 0}) \qquad fx \approx gy \to fy \approx 0.$$

PROOF. First we note that, in the presence of $\varepsilon_{\mathbf{R}_1, x \equiv y}$, $\varepsilon_{\mathbf{R}_8, fy \equiv 0}$ is equivalent to $fx \approx gy \to fx \approx y$. For assume that, in an algebra satisfying $\varepsilon_{\mathbf{R}_1, x \equiv y}$, we have elements x, y such that $f(x) = g(y)$. If $f(x) = y$, then $f^2(x) = f(y) = 0$. On the other hand, if $f(y) = 0$, then $f(y) = f(f(x))$ and $g(y) = g(g(y)) = g(f(x))$, whence $y = f(x)$ by $\varepsilon_{\mathbf{R}_1, x \equiv y}$.

Now let $\mathbf{S}$ be any algebra in $\mathcal{M}$ satisfying our quasi-equations. We want to show that $\mathbf{S}$ is a subdirect product of copies of $\mathbf{T}_1$ and $\mathcal{F}_1$. For any nonzero $x \in \mathbf{S}$, $\mathrm{Sg}(x) \cong \mathbf{T}_1$ or $\mathcal{F}_1$, as the other 1-generated algebras fail one quasi-equation or the other.

This enables us to identify a minimal generating set A for $\mathbf{S}$, namely, $A = \{a \in S : a = g(b) \text{ implies } a = b\}$. Moreover, if $a \in A$ and $\mathrm{Sg}(a) \cong \mathbf{T}_1$, then $\mathbf{S} = \{0, a\} * (\mathbf{S} \setminus \{a\})$. Thus without loss of generality we may assume that $\mathrm{Sg}(a) \cong \mathcal{F}_1$ for all $a \in A$. It remains to figure out how these subalgebras fit together.

Let $B = \{f(a) : a \in A\}$ and $C = \{g(a) : a \in A\}$. Using the equations defining $\mathcal{M}$, we see that $S = A \cup B \cup C \cup \{0\}$. Moreover, applying the quasi-equation $fx \approx gy \to fx \approx y$ and the assumptions of the previous paragraph, this is a disjoint union.

Now we define a collection of congruences on $\mathbf{S}$.

- Let φ be the equivalence whose blocks are A, B, C, $\{0\}$. Clearly this is a congruence with $\mathbf{S}/\varphi \cong \mathcal{F}_1$.
- For $b \in B$, let σ_b be the equivalence relation whose blocks are $\{a \in A : f(a) = b\}$, $\{b\}$, $\{a' \in A : f(a') \neq b\} \cup C$, $B \setminus \{b\} \cup \{0\}$. Again, for each $b \in B$ this is a congruence with $\mathbf{S}/\sigma_b \cong \mathcal{F}_1$.
- For $c \in C$, let τ_c be the equivalence relation whose blocks are $\{c\} \cup \{a \in A : g(a) = c\}$ and the rest of S. Each τ_c is likewise a congruence with $\mathbf{S}/\tau_c \cong \mathbf{T}_1$.

Now it remains only to observe that, in view of $\varepsilon_{\mathbf{R}_1, x \equiv y}$,

$$\varphi \wedge \bigwedge_{b \in B} \sigma_b \wedge \bigwedge_{c \in C} \tau_c = \Delta$$

as desired. $\qquad\qquad\qquad\qquad\qquad\qquad\qquad\qquad\qquad\qquad\qquad\qquad\square$

PROPOSITION 5.12. *The algebra $\mathcal{F}_1$ satisfies the following:*

(1) $\langle \mathcal{F}_1 \rangle$ *has the quasi-equational basis*

$$(\varepsilon_{\mathbf{R}_1, x \equiv y}) \qquad fx \approx fy \ \& \ gx \approx gy \to x \approx y,$$
$$(\varepsilon_{\mathbf{R}_8, fy \equiv 0}) \qquad fx \approx gy \to fy \approx 0.$$

(2) *The only quasicritical algebras in $\langle \mathcal{F}_1 \rangle$ are $\mathbf{T}_1$ and $\mathcal{F}_1$.*
(3) $\mathfrak{Q} \leq \langle \mathcal{F}_1 \rangle$ *if and only if $\mathfrak{Q}$ omits $\mathbf{T}_2$, $\mathbf{T}_5$, $\mathbf{R}_1$, and $\mathbf{R}_8$.*
(4) $\mathbf{S} \in \langle \mathcal{F}_1 \rangle$ *if and only if $\mathbf{S}$ omits $\mathbf{T}_2$, $\mathbf{T}_3$, $\mathbf{T}_4$, $\mathbf{T}_5$, $\mathbf{R}_1$, $\mathbf{R}_8$, and $\mathbf{P}_2$.*
(5) $\langle \mathcal{F}_1 \rangle_* = \langle \mathbf{T}_1 \rangle$.
(6) *The reflection congruences of $\mathcal{F}_1$ are $\Delta, \nabla, fx \equiv 0$,* $(fx \equiv 0) \vee (gx \equiv x)$, $fx \equiv gx \equiv 0$.
(7) $\gamma^{\mathcal{F}_1}(fx \equiv 0) = \rho^{\mathcal{F}_1}_{(fx \approx 0)}$ *is $fx \equiv 0$,* $\gamma^{\mathcal{F}_1}(fx \equiv gx) = \gamma^{\mathcal{F}_1}(gx \equiv 0)$ *is $fx \equiv gx \equiv 0$.*
(8) $\mathrm{diag}(\mathcal{F}_1)$ *is $x \approx x$.*
(9) $\mathcal{E}(\mathcal{F}_1) = \{\varepsilon_{\mathcal{F}_1, fx \equiv 0}, \varepsilon_{\mathcal{F}_1, gx \equiv 0}, \varepsilon_{\mathcal{F}_1, fx \equiv gx}\}$ $= \{fx \approx 0, gx \approx 0, fx \approx gx\}$.
(10) $\kappa(\mathcal{F}_1) = \{\widehat{\mathcal{U}}(\mathcal{F}_1, fx \equiv 0)\}$.
(11) $\widehat{\mathcal{U}}(\mathcal{F}_1, fx \equiv 0) = \langle fx \approx 0 \rangle = \langle \mathbf{T}_1 \rangle \vee \langle \mathbf{T}_5 \rangle$, $\mathcal{U}(\mathcal{F}_1, fx \equiv gx) = \mathcal{U}(\mathcal{F}_1, gx \equiv 0) = \langle fx \approx gx \rangle = \langle \mathbf{T}_5 \rangle$.
(12) $\mathbf{S} \in \mathcal{U}(\mathcal{F}_1, fx \equiv gx)$ *if and only if $\mathbf{S}$ omits $\mathbf{T}_1$,* $\mathbf{S} \in \widehat{\mathcal{U}}(\mathcal{F}_1, fx \equiv 0)$ *if and only if $\mathbf{S}$ omits $\mathbf{T}_3$, $\mathbf{T}_4$ and $\mathcal{F}_1$.*
(13) *See Figure 5.9 for the segment of $\mathrm{L_q}(\mathcal{M})$ related to $\mathcal{F}_1$.*

PROOF. The algebra $\mathcal{F}_1$ has subalgebras isomorphic to $\mathbf{T}_0$, $\mathbf{T}_1$, and $\mathbf{T}_1 * \mathbf{T}_1$. The congruence lattice of $\mathcal{F}_1$ is given in Figure 5.4; note that $fx \equiv 0$ is its unique minimal nonzero reflection congruence. The equations in (9) reflect the discussion of Section 4.3. Claims (2) and (11) use (10) and the characterizations of $\langle fx \approx 0 \rangle$ in Lemma 5.3 and $\langle \mathbf{T}_5 \rangle$ in Lemma 5.5.   □

Thus the inclusions in Figure 5.5 below $\langle \mathbf{T}_5 \rangle$ and $\langle \mathcal{F}_1 \rangle$ are covers. As we shall see later, this is also true for $\langle \mathbf{R}_1 \rangle$ and $\langle \mathbf{R}_8 \rangle$.

In the process of analyzing the lattice of subquasivarieties $L_q(\mathcal{M})$, we have inadvertently found the lattice of subvarieties $L_v(\mathcal{M})$. The crucial facts are that $\mathbb{V}(\mathcal{F}_1) = \mathcal{M}$ (since $\mathcal{F}_n = \mathcal{F}_1 * \cdots * \mathcal{F}_1$), and $\kappa(\mathcal{F}_1) = \langle \mathbf{T}_1 \rangle \vee \langle \mathbf{T}_5 \rangle = \langle fx \approx 0 \rangle$. The subvariety lattice is given in Figure 5.7.

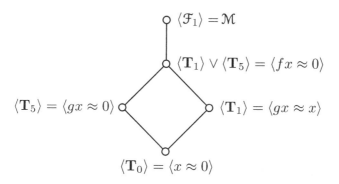

FIGURE 5.7. The lattice of subvarieties $L_v(\mathcal{M})$

5.3. Illustrating the Algorithms: 2-Generated Algebras

Now we turn to the 2-generated small quasicritical algebras in $\mathcal{M}$, except $\mathbf{P}_2$, which will be postponed to the next section, along with $\mathbf{T}_3$. Recall that $\langle \mathbf{T}_4 \rangle$, $\langle \mathbf{R}_3 \rangle$, and $\langle \mathbf{R}_5 \rangle$ are not finitely based. For the 1-generated quasicritical algebras except $\mathbf{T}_4$, we were able to find a basis for $\langle \mathbf{T} \rangle$ consisting of quasi-equations in at most 2 variables. As per the discussion following Theorem 5.2, $\langle \mathbf{R}_6 \rangle$ also has a 2-variable basis. Likewise, $\langle \mathbf{R}_2 \rangle$ and $\langle \mathbf{R}_4 \rangle$ both have a finite basis, but we don't know the number of variables. (A bound is given in [43].) It is unknown whether $\langle \mathbf{R}_1 \rangle$, $\langle \mathbf{R}_7 \rangle$, $\langle \mathbf{R}_8 \rangle$, or $\langle \mathbf{P}_2 \rangle$ is finitely based. In all the cases where we do not know the quasi-equational basis, the statements of parts (1)–(4) of the analysis are necessarily weaker!

First we consider $\mathbf{R}_1$. The congruence lattice of $\mathbf{R}_1$ is drawn in Figure 5.8.

PROPOSITION 5.13. *The algebra* $\mathbf{R}_1$ *satisfies the following:*

(1) $\langle \mathbf{R}_1 \rangle$ *satisfies the quasi-equations*

$$(\varepsilon_{\mathbf{T}_2, gx \equiv x}) \qquad fx \approx 0 \to gx \approx x,$$
$$(\varepsilon_{\mathbf{R}_8, fy \equiv 0}) \qquad fx \approx gy \to fy \approx 0.$$

(2) *The only quasicritical algebras in* $\langle \mathbf{R}_1 \rangle$ *are* $\mathbf{T}_1$, $\mathcal{F}_1$, *and* $\mathbf{R}_1$.
(3) *If* $\mathcal{Q} \le \langle \mathbf{R}_1 \rangle$, *then* $\mathcal{Q}$ *omits* $\mathbf{T}_2$, $\mathbf{T}_5$, *and* $\mathbf{R}_8$.
(4) *If* $\mathbf{S} \in \langle \mathbf{R}_1 \rangle$, *then* $\mathbf{S}$ *omits* $\mathbf{T}_2$, $\mathbf{T}_3$, $\mathbf{T}_4$, $\mathbf{T}_5$, $\mathbf{P}_2$, *and* $\mathbf{R}_8$.

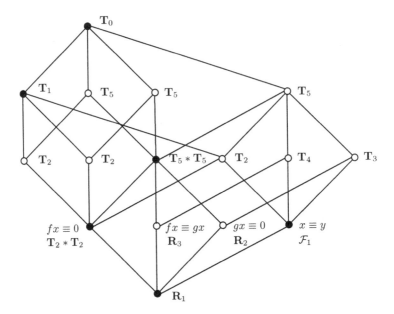

FIGURE 5.8. Con $\mathbf{R}_1$ where $\mathbf{R}_1 = \mathcal{F}_2/\theta$ for the congruence $\theta = (fx \equiv fy) \vee (gx \equiv gy)$. Here $\mathbf{T}_2 * \mathbf{T}_2$ denotes the non-quasicritical algebra consisting of two copies of $\mathbf{T}_2$ glued together.

(5) $\langle \mathbf{R}_1 \rangle_* = \langle \mathbf{R}_1 \rangle \wedge \widehat{\mathcal{U}}(\mathbf{R}_1, x \equiv y) = \langle \mathcal{F}_1 \rangle$.
(6) *The reflection congruences of* $\mathbf{R}_1$ *are* Δ, ∇, $fx \equiv 0$, $x \equiv y$, $(fx \equiv 0) \vee (gx \equiv 0)$, $(gx \equiv x) \vee (x \equiv y)$.
(7) $\gamma^{\mathbf{R}_1}(fx \equiv 0)$ *is* $fx \equiv 0$,
   $\gamma^{\mathbf{R}_1}(fx \equiv gx) = \gamma^{\mathbf{R}_1}(gx \equiv 0)$ *is* $fx \equiv gx \equiv 0$,
   $\gamma^{\mathbf{R}_1}(x \equiv y)$ *is* $x \equiv y$.
(8) $\mathrm{diag}(\mathbf{R}_1)$ *is* $fx \approx fy$ & $gx \approx gy$.
(9) $\mathcal{E}(\mathbf{R}_1) = \{\varepsilon_{\mathbf{R}_1, fx \equiv gx}, \varepsilon_{\mathbf{R}_1, fx \equiv 0}, \varepsilon_{\mathbf{R}_1, gx \equiv 0}, \varepsilon_{\mathbf{R}_1, x \equiv y}\}$
   $= \{fx \approx gx, fx \approx 0, gx \approx 0, \varepsilon_{\mathbf{R}_1, x \equiv y}\}$.
(10) $\kappa(\mathbf{R}_1) = \{\widehat{\mathcal{U}}(\mathbf{R}_1, fx \equiv 0), \widehat{\mathcal{U}}(\mathbf{R}_1, x \equiv y)\}$.
(11) $\mathcal{U}(\mathbf{R}_1, fx \equiv gx) = \mathcal{U}(\mathbf{R}_1, gx \equiv 0) = \langle \mathbf{T}_5 \rangle$,
   $\widehat{\mathcal{U}}(\mathbf{R}_1, fx \equiv 0) = \langle fx \approx 0 \rangle = \langle \mathbf{T}_1 \rangle \vee \langle \mathbf{T}_5 \rangle$,
   $\widehat{\mathcal{U}}(\mathbf{R}_1, x \equiv y) = \langle fx \approx fy$ & $gx \approx gy \rightarrow x \approx y \rangle = \langle \mathbf{R}_6 \rangle$.
(12) $\mathbf{S} \in \mathcal{U}(\mathbf{R}_1, gx \equiv 0)$ *if and only if* $\mathbf{S}$ *omits* $\mathbf{T}_1$,
   $\mathbf{S} \in \widehat{\mathcal{U}}(\mathbf{R}_1, fx \equiv 0)$ *if and only if* $\mathbf{S}$ *omits* $\mathbf{T}_3$, $\mathbf{T}_4$, *and* $\mathcal{F}_1$,
   $\mathbf{S} \in \widehat{\mathcal{U}}(\mathbf{R}_1, x \equiv y)$ *if and only if* $\mathbf{S}$ *omits* $\mathbf{T}_2$, $\mathbf{T}_5$, $\mathbf{R}_1$, $\mathbf{R}_2$, *and* $\mathbf{R}_3$.
(13) *See* Figure 5.9 *for the segment of* $\mathrm{L}_q(\mathcal{M})$ *related to* $\mathbf{R}_1$.

PROOF. First note that $\mathcal{F}_1 \leq \mathbf{R}_1$. The inclusions $\langle \mathbf{R}_1 \rangle \leq \langle \mathbf{R}_j \rangle$ for $j = 2, 3, 7$ are explained in Table 5.3. There are also join dependencies

$\langle \mathbf{R}_1 \rangle \leq \langle \mathbf{T}_2 \rangle \vee \langle \mathcal{F}_1 \rangle < \langle \mathbf{T}_5 \rangle \vee \langle \mathcal{F}_1 \rangle$, using $\mathbf{T}_2 \leq \mathbf{T}_1 \times \mathbf{T}_5$ and that $\langle \mathbf{T}_5 \rangle$ is join prime.

The quasi-equations $\varepsilon_{\mathbf{R}_1, fx \equiv 0}$, $\varepsilon_{\mathbf{R}_1, fx \equiv gx}$, and $\varepsilon_{\mathbf{R}_1, gx \equiv 0}$ are logically equivalent to the equations $fx \approx 0$, $fx \approx gx$, and $gx \approx 0$, respectively.

To see that $\varepsilon_{\mathbf{R}_1, fx \equiv 0}$, that is $fx \approx fy \ \& \ gx \approx gy \rightarrow fx \approx 0$, is equivalent to $fx \approx 0$, first note the equation $fx \approx 0$ implies any quasi-equation whose conclusion is $fx \approx 0$. Conversely, consider the substitution $y \mapsto x$ in $\varepsilon_{\mathbf{R}_1, fx \equiv 0}$. The hypothesis of the resulting quasi-equation, $fx \approx fx \ \& \ gx \approx gx$, is always true, while the conclusion is $fx \approx 0$. Similar arguments show that $\varepsilon_{\mathbf{R}_1, gx \equiv 0}$ and $\varepsilon_{\mathbf{R}_1, fx \equiv gx}$ are equivalent to their conclusions.

Moreover, as $\mathbf{R}_1$ satisfies $\varepsilon_{\mathbf{R}_8, fy \equiv 0}$, the meet $\langle \mathbf{R}_1 \rangle_* = \langle \mathbf{R}_1 \rangle \wedge \widehat{\mathcal{U}}(\mathbf{R}_1, x \equiv y)$ satisfies both $\varepsilon_{\mathbf{R}_8, fy \equiv 0}$ and $\varepsilon_{\mathbf{R}_1, x \equiv y}$, whence $\langle \mathbf{R}_1 \rangle_* = \langle \mathcal{F}_1 \rangle$ by Lemma 5.11. The proof that $\widehat{\mathcal{U}}(\mathbf{R}_1, x \equiv y) = \langle \mathbf{R}_6 \rangle$ will be given in Proposition 5.20. □

COROLLARY 5.14. *A quasivariety* $\mathcal{Q} \leq \mathcal{M}$ *satisfies*

$$fx \approx fy \ \& \ gx \approx gy \rightarrow x \approx y$$

if and only if $\mathcal{Q}$ *does not contain* $\mathbf{T}_2$, $\mathbf{T}_5$, *and* $\mathbf{R}_1$.

We do not know whether $\langle \mathbf{R}_1 \rangle$ is finitely based. In particular, *Are the quasi-equations* $\varepsilon_{\mathbf{T}_2, gx \equiv x}$ *and* $\varepsilon_{\mathbf{R}_8, fy \equiv 0}$ *in* (1) *a basis for* $\langle \mathbf{R}_1 \rangle$?

The next lemma summarizes some of the equalities of quasivarieties that have been developed above.

LEMMA 5.15. *The following hold in* $L_q(\mathcal{M})$.
  (1) $\langle \mathbf{T}_1 \rangle = \langle fx \approx 0 \ \& \ gx \approx x \rangle$.
  (2) $\mathcal{U}(\mathbf{R}_1, gx \equiv 0) = \mathcal{U}(\mathbf{R}_1, fx \equiv gx) = \langle gx \approx 0 \rangle = \langle \mathbf{T}_5 \rangle$.
  (3) $\widehat{\mathcal{U}}(\mathbf{R}_1, fx \equiv 0) = \langle \mathbf{T}_1 \rangle \vee \langle \mathbf{T}_5 \rangle$.
  (4) $\widehat{\mathcal{U}}(\mathbf{R}_1, fx \equiv 0) \wedge \widehat{\mathcal{U}}(\mathbf{R}_1, x \equiv y) = \langle \mathbf{T}_1 \rangle$.
  (5) $\widehat{\mathcal{U}}(\mathbf{R}_1, fx \equiv 0) \vee \langle \mathbf{R}_1 \rangle = \langle \mathbf{T}_5 \rangle \vee \langle \mathcal{F}_1 \rangle$.

PROOF. The equality (1) holds as the only quasicritical that satisfies $fx \approx 0$ and $gx \approx x$ is $\mathbf{T}_1$.

Equations (2) and (3) use Lemmas 5.3 and 5.5.

Neither $\mathbf{T}_2$ nor $\mathbf{T}_5$ satisfies $\varepsilon_{\mathbf{R}_1, x \equiv y}$ but $\mathbf{T}_1$ does, giving (4).

Finally, $\langle \mathcal{F}_1 \rangle \leq \langle \mathbf{R}_1 \rangle$ but $\mathbf{R}_1 \leq \mathbf{T}_5 \times \mathcal{F}_1$, implying equation (5). □

Figure 5.9 illustrates Lemma 5.15 and the structure of the lower segment in $L_q(\mathcal{M})$.

The congruence lattices of $\mathbf{R}_2$, $\mathbf{R}_3$, $\mathbf{R}_4$, and $\mathbf{R}_5$ are all isomorphic as lattices. The labelings of the congruence lattices only change at the atoms and the least element; see Figure 5.10 for two of them.

  • For $\mathbf{R}_2$ the atoms are $fx \equiv 0$ with quotient $\mathbf{T}_5 * \mathbf{T}_5$, and $x \equiv y$ with quotient $\mathbf{T}_3$.
  • For $\mathbf{R}_3$ the atoms are $fx \equiv 0$ with quotient $\mathbf{T}_5 * \mathbf{T}_5$, and $x \equiv y$ with quotient $\mathbf{T}_4$.

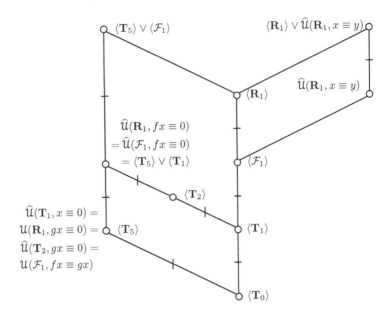

FIGURE 5.9. Segment of $L_q(\mathcal{M})$ related to $\langle \mathbf{T}_1 \rangle$, $\langle \mathbf{T}_2 \rangle$, $\langle \mathbf{T}_5 \rangle$, $\langle \mathcal{F}_1 \rangle$, and $\langle \mathbf{R}_1 \rangle$. Ticks indicate covers.

- For $\mathbf{R}_4$ the atoms are $gx \equiv 0$ with quotient $\mathbf{T}_5 * \mathbf{T}_5$, and $gx \equiv x$ with quotient $\mathbf{T}_3$.
- For $\mathbf{R}_5$ the atoms are $gy \equiv 0$ with quotient $\mathbf{T}_5 * \mathbf{T}_5$, and $gy \equiv y$ with quotient $\mathbf{T}_4$.

PROPOSITION 5.16. *The algebra* $\mathbf{R}_2$ *satisfies the following:*

(1) $\langle \mathbf{R}_2 \rangle$ *is finitely based and satisfies*

$$(\varepsilon_{\mathbf{T}_2, gx \equiv x}) \qquad fx \approx 0 \rightarrow gx \approx x,$$

$$(\varepsilon_{\mathbf{T}_4, fx \equiv 0}) \qquad fx \approx gx \rightarrow fx \approx 0.$$

(2) $\mathbf{T}_3, \mathbf{P}_2, \mathbf{R}_8, \mathbf{R}_1, \mathcal{F}_1, \mathbf{T}_1 \in \langle \mathbf{R}_2 \rangle$.

(3) *If* $\mathfrak{Q} \leq \langle \mathbf{R}_2 \rangle$, *then* $\mathfrak{Q}$ *omits* $\mathbf{T}_2$, $\mathbf{T}_4$, *and* $\mathbf{T}_5$.

(4) *If* $\mathbf{S} \in \langle \mathbf{R}_2 \rangle$, *then* $\mathbf{S}$ *omits* $\mathbf{T}_2$, $\mathbf{T}_4$, *and* $\mathbf{T}_5$.

(5) $\langle \mathbf{R}_2 \rangle_* = \langle \mathbf{R}_2 \rangle \wedge \widehat{\mathcal{U}}(\mathbf{R}_2, x \equiv y) \geq \langle \mathbf{T}_3 \rangle \vee \langle \mathbf{R}_1 \rangle$.

(6) *The reflection congruences of* $\mathbf{R}_2$ *are* Δ, ∇, $fx \equiv 0$, $x \equiv y$.

(7) $\gamma^{\mathbf{R}_2}(fx \equiv 0) = \rho^{\mathbf{R}_2}_{\langle fx \approx 0 \rangle}$ *is* $fx \equiv 0$,

$\gamma^{\mathbf{R}_2}(x \equiv y) = \rho^{\mathbf{R}_2}_{\langle x \approx y \rangle}$ *is* $x \equiv y$.

(8) $\mathrm{diag}(\mathbf{R}_2)$ *is* $fx \approx fy \;\&\; gx \approx gy \approx 0$.

(9) $\mathcal{E}(\mathbf{R}_2) = \{\varepsilon_{\mathbf{R}_2, fx \equiv 0}, \varepsilon_{\mathbf{R}_2, x \equiv y}\}$.

(10) $\kappa(\mathbf{R}_2) = \lambda(\mathbf{R}_2) = \{\widehat{\mathcal{U}}(\mathbf{R}_2, fx \equiv 0), \widehat{\mathcal{U}}(\mathbf{R}_2, x \equiv y)\}$.

(11) $\widehat{\mathcal{U}}(\mathbf{R}_2, fx \equiv 0) = \langle gx \approx 0 \rightarrow fx \approx 0 \rangle$,

$\langle \mathbf{T}_5 \rangle \vee \langle \mathbf{R}_3 \rangle \vee \langle \mathbf{R}_5 \rangle \leq \widehat{\mathcal{U}}(\mathbf{R}_2, fx \equiv 0)$ *but* $\langle \mathbf{T}_3 \rangle \not\leq \widehat{\mathcal{U}}(\mathbf{R}_2, fx \equiv 0)$,

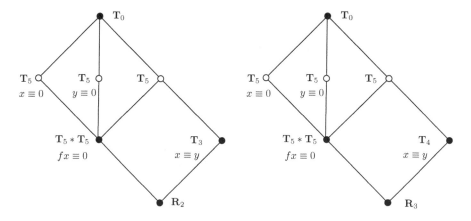

FIGURE 5.10. Congruence lattices of $\mathbf{R}_2$ and $\mathbf{R}_3$

$$\langle \mathbf{R}_3 \rangle \vee \langle \mathbf{R}_4 \rangle \vee \langle \mathbf{R}_5 \rangle \vee \langle \mathbf{R}_6 \rangle \leq \widehat{\mathcal{U}}(\mathbf{R}_2, x \equiv y) \text{ but } \langle \mathbf{T}_5 \rangle, \langle \mathbf{R}_2 \rangle \nleq \widehat{\mathcal{U}}(\mathbf{R}_2, x \equiv y).$$

(12) $\mathbf{S} \in \widehat{\mathcal{U}}(\mathbf{R}_2, fx \equiv 0)$ *if and only if* $\mathbf{S}$ *omits* $\mathbf{T}_3$,
$\mathbf{S} \in \widehat{\mathcal{U}}(\mathbf{R}_2, x \equiv y)$ *if and only if* $\mathbf{S}$ *omits* $\mathbf{T}_5$ *and* $\mathbf{R}_2$.

PROOF. The congruence lattice of $\mathbf{R}_2$ is in Figure 5.10. The quasivariety $\langle \mathbf{R}_2 \rangle$ is finitely based by Theorem 5.2, but we do not know whether the quasi-equations in (1) form a basis. The crucial observations are that $\mathbf{R}_1 \leq \mathbf{R}_2 \times \mathbf{T}_1 \leq \mathbf{R}_2^2$, while $\mathbf{T}_3 \leq \mathbf{R}_2$. The substitution $y \mapsto x$ can be used to derive the equivalence $\widehat{\mathcal{U}}(\mathbf{R}_2, fx \equiv 0) = \langle gx \approx 0 \to fx \approx 0 \rangle$. $\qquad\square$

PROPOSITION 5.17. *The algebra* $\mathbf{R}_3$ *satisfies the following:*

(1) $\langle \mathbf{R}_3 \rangle$ *is not finitely based. Its quasi-identities include*

$$(\varepsilon_{\mathbf{T}_2, gx \equiv x}) \qquad fx \approx 0 \to gx \approx x,$$

$$(\varepsilon_{\mathbf{P}_2, fx \equiv gx}) \qquad fx \approx gy \ \& \ gx \approx fy \to fx \approx gx,$$

and more generally the quasi-equations $\varepsilon_{\mathbf{P}_n, fx_1 \equiv gx_1}$ *for* $n \geq 2$ *of Theorem 5.26.*

(2) $\mathbf{T}_1, \mathbf{T}_4, \mathbf{R}_8, \mathcal{F}_1, \mathbf{R}_1 \in \langle \mathbf{R}_3 \rangle$.

(3) *If* $\mathcal{Q} \leq \langle \mathbf{R}_3 \rangle$, *then* $\mathcal{Q}$ *omits* $\mathbf{T}_2$, $\mathbf{T}_5$, *and* $\mathbf{P}_2$.

(4) *If* $\mathbf{S} \in \langle \mathbf{R}_3 \rangle$, *then* $\mathbf{S}$ *omits* $\mathbf{T}_2$, $\mathbf{T}_3$, $\mathbf{T}_5$, *and* $\mathbf{P}_2$.

(5) $\langle \mathbf{R}_3 \rangle_* = \langle \mathbf{R}_3 \rangle \wedge \widehat{\mathcal{U}}(\mathbf{R}_3, x \equiv y) \geq \langle \mathbf{T}_4 \rangle \vee \langle \mathbf{R}_1 \rangle$.

(6) *The reflection congruences of* $\mathbf{R}_3$ *are* $\Delta, \nabla, fx \equiv 0, x \equiv y$.

(7) $\gamma^{\mathbf{R}_3}(fx \equiv 0) = \rho^{\mathbf{R}_3}_{\langle fx \approx 0 \rangle}$ *is* $fx \equiv 0$,
$\gamma^{\mathbf{R}_3}(x \equiv y) = \rho^{\mathbf{R}_3}_{\langle x \approx y \rangle}$ *is* $x \equiv y$.

(8) $\text{diag}(\mathbf{R}_3)$ *is* $fy \approx gy \approx fx \approx gx$.

(9) $\mathcal{E}(\mathbf{R}_3) = \{\varepsilon_{\mathbf{R}_3, fx \equiv 0}, \varepsilon_{\mathbf{R}_3, x \equiv y}\}$.

(10) $\kappa(\mathbf{R}_3) = \lambda(\mathbf{R}_3) = \{\widehat{\mathcal{U}}(\mathbf{R}_3, fx \equiv 0), \widehat{\mathcal{U}}(\mathbf{R}_3, x \equiv y)\}$.

(11) $\widehat{\mathcal{U}}(\mathbf{R}_3, fx \equiv 0) = \langle fx \approx gx \to fx \approx 0 \rangle$,

$\langle \mathbf{T}_5 \rangle \vee \langle \mathbf{R}_2 \rangle \vee \langle \mathbf{R}_4 \rangle \leq \widehat{\mathcal{U}}(\mathbf{R}_3, fx \equiv 0)$ *but* $\langle \mathbf{T}_4 \rangle \not\leq \widehat{\mathcal{U}}(\mathbf{R}_3, fx \equiv 0)$,

$\langle \mathbf{R}_2 \rangle \vee \langle \mathbf{R}_4 \rangle \vee \langle \mathbf{R}_5 \rangle \vee \langle \mathbf{R}_6 \rangle \leq \widehat{\mathcal{U}}(\mathbf{R}_3, x \equiv y)$ *but* $\langle \mathbf{T}_5 \rangle, \langle \mathbf{R}_3 \rangle \not\leq$ $\widehat{\mathcal{U}}(\mathbf{R}_3, x \equiv y)$.

(12) $\mathbf{S} \in \widehat{\mathcal{U}}(\mathbf{R}_3, fx \equiv 0)$ *if and only if* $\mathbf{S}$ *omits* $\mathbf{T}_4$,

$\mathbf{S} \in \widehat{\mathcal{U}}(\mathbf{R}_3, x \equiv y)$ *if and only if* $\mathbf{S}$ *omits* $\mathbf{T}_5$ *and* $\mathbf{R}_3$.

PROOF. The congruence lattice of $\mathbf{R}_3$ is in Figure 5.10. The quasivariety $\langle \mathbf{R}_3 \rangle$ is not finitely based by Theorem 5.2. Also observe that $\mathbf{R}_1 \leq \mathbf{R}_3 \times \mathbf{T}_1 \leq \mathbf{R}_3{}^2$ and $\mathbf{T}_4 \leq \mathbf{R}_3$. The equivalence $\widehat{\mathcal{U}}(\mathbf{R}_3, fx \equiv 0) = \langle fx \approx gx \to fx \approx 0 \rangle$ follows from the substitution $y \mapsto x$. $\qquad \square$

PROPOSITION 5.18. *The algebra* $\mathbf{R}_4$ *satisfies the following.*

(1) $\langle \mathbf{R}_4 \rangle$ *is finitely based and satisfies*

$$(\varepsilon_{\mathbf{T}_5, x \equiv 0}) \qquad fx \approx gx \approx 0 \to x \approx 0,$$

$$(\varepsilon_{\mathbf{T}_4, fx \equiv 0}) \qquad fx \approx gx \to fx \approx 0.$$

(2) $\mathbf{T}_1, \mathbf{T}_2, \mathbf{T}_3, \mathbf{R}_1, \mathbf{P}_2, \mathcal{F}_1, \mathbf{R}_7, \mathbf{R}_8 \in \langle \mathbf{R}_4 \rangle$.

(3) *If* $\mathcal{Q} \leq \langle \mathbf{R}_4 \rangle$, *then* $\mathcal{Q}$ *omits* $\mathbf{T}_4$, $\mathbf{T}_5$, *and* $\mathbf{R}_2$.

(4) *If* $\mathbf{S} \in \langle \mathbf{R}_4 \rangle$, *then* $\mathbf{S}$ *omits* $\mathbf{T}_4$, $\mathbf{T}_5$, *and* $\mathbf{R}_2$.

(5) $\langle \mathbf{R}_4 \rangle_* = \langle \mathbf{R}_4 \rangle \wedge \widehat{\mathcal{U}}(\mathbf{R}_4, gx \equiv x) \geq \langle \mathbf{T}_3 \rangle \vee \langle \mathbf{R}_7 \rangle$.

(6) *The reflection congruences of* $\mathbf{R}_4$ *are* $\Delta, \nabla, gx \equiv 0, gx \equiv x$.

(7) $\gamma^{\mathbf{R}_4}(gx \equiv 0) = \rho^{\mathbf{R}_4}_{\langle gx \approx 0 \rangle}$ *is* $gx \equiv 0$,

$\gamma^{\mathbf{R}_4}(gx \equiv x) = \rho^{\mathbf{R}_4}_{\langle gx \approx x \rangle}$ *is* $gx \equiv x$.

(8) $\mathrm{diag}(\mathbf{R}_4)$ *is* $fx \approx gy \approx 0$ & $gx \approx fy$.

(9) $\mathcal{E}(\mathbf{R}_4) = \{\varepsilon_{\mathbf{R}_4, gx \equiv 0}, \varepsilon_{\mathbf{R}_4, gx \equiv x}\}$.

(10) $\kappa(\mathbf{R}_4) = \lambda(\mathbf{R}_4) = \{\widehat{\mathcal{U}}(\mathbf{R}_4, gx \equiv 0), \widehat{\mathcal{U}}(\mathbf{R}_4, gx \equiv x)\}$.

(11) $\widehat{\mathcal{U}}(\mathbf{R}_4, gx \equiv 0) = \langle gx \approx 0 \to fx \approx 0 \rangle$,

$\langle \mathbf{T}_5 \rangle \vee \langle \mathbf{R}_3 \rangle \vee \langle \mathbf{R}_5 \rangle \leq \widehat{\mathcal{U}}(\mathbf{R}_4, gx \equiv 0)$ *but* $\langle \mathbf{T}_3 \rangle \not\leq \widehat{\mathcal{U}}(\mathbf{R}_4, gx \equiv 0)$,

$\langle \mathbf{R}_2 \rangle \vee \langle \mathbf{R}_3 \rangle \vee \langle \mathbf{R}_5 \rangle \vee \langle \mathbf{R}_6 \rangle \leq \widehat{\mathcal{U}}(\mathbf{R}_4, gx \equiv x)$ *but* $\langle \mathbf{T}_5 \rangle, \langle \mathbf{R}_4 \rangle \not\leq$ $\widehat{\mathcal{U}}(\mathbf{R}_4, gx \equiv x)$.

(12) $\mathbf{S} \in \widehat{\mathcal{U}}(\mathbf{R}_4, gx \equiv 0)$ *if and only if* $\mathbf{S}$ *omits* $\mathbf{T}_3$,

$\mathbf{S} \in \widehat{\mathcal{U}}(\mathbf{R}_4, gx \equiv x)$ *if and only if* $\mathbf{S}$ *omits* $\mathbf{T}_5$ *and* $\mathbf{R}_4$.

PROOF. The quasivariety $\langle \mathbf{R}_4 \rangle$ is finitely based by Theorem 5.2, but we do not know whether the quasi-equations in (1) form a basis. Note that $\mathbf{R}_7 \leq \mathbf{R}_4 \times \mathbf{T}_1 \leq \mathbf{R}_4^2$ and $\mathbf{T}_3 \leq \mathbf{R}_4$.

The argument that $\widehat{\mathcal{U}}(\mathbf{R}_4, gx \equiv 0) = \langle gx \approx 0 \to fx \approx 0 \rangle$ is tricky. For one direction use the substitution $x \mapsto fx$ and $y \mapsto x$; for the other use, $x \mapsto gx$ and $y \mapsto x$. We leave the details to the reader, noting that we don't need them anyway: both quasivarieties are characterized by omitting $\mathbf{T}_3$ as a subalgebra! $\qquad \square$

PROPOSITION 5.19. *The algebra* $\mathbf{R}_5$ *satisfies the following:*

(1) $\langle \mathbf{R}_5 \rangle$ *is not finitely based. Its quasi-identities include*

$$(\varepsilon_{\mathbf{T}_5, x \equiv 0}) \qquad fx \approx gx \approx 0 \to x \approx 0,$$

$$(\varepsilon_{\mathbf{P}_2, fx \equiv gx}) \qquad fx \approx gy \ \& \ gx \approx fy \to fx \approx gx,$$

and more generally the quasi-equations $\varepsilon_{\mathbf{P}_n, fx_1 \equiv gx_1}$ *for* $n \geq 2$ *of Theorem 5.26.*

(2) $\mathbf{T}_1, \mathbf{T}_2, \mathbf{T}_4, \mathbf{R}_1, \mathcal{F}_1, \mathbf{R}_7, \mathbf{R}_8 \in \langle \mathbf{R}_5 \rangle$.

(3) *If* $\mathcal{Q} \leq \langle \mathbf{R}_5 \rangle$, *then* $\mathcal{Q}$ *omits* $\mathbf{T}_5$, $\mathbf{R}_3$, *and* $\mathbf{P}_2$.

(4) *If* $\mathbf{S} \in \langle \mathbf{R}_5 \rangle$, *then* $\mathbf{S}$ *omits* $\mathbf{T}_5$, $\mathbf{T}_3$, $\mathbf{R}_3$, *and* $\mathbf{P}_2$.

(5) $\langle \mathbf{R}_5 \rangle_* = \langle \mathbf{R}_5 \rangle \wedge \widehat{\mathcal{U}}(\mathbf{R}_5, gy \equiv y) \geq \langle \mathbf{T}_4 \rangle \vee \langle \mathbf{R}_7 \rangle$.

(6) *The reflection congruences of* $\mathbf{R}_5$ *are* $\Delta, \nabla, gy \equiv 0, gy \equiv y$.

(7) $\gamma^{\mathbf{R}_5}(gy \equiv 0) = \rho^{\mathbf{R}_5}_{\langle gy \approx 0 \rangle}$ *is* $gy \equiv 0$,

$\gamma^{\mathbf{R}_5}(gy \equiv y) = \rho^{\mathbf{R}_5}_{\langle gy \approx y \rangle}$ *is* $gy \equiv y$.

(8) $\mathrm{diag}(\mathbf{R}_5)$ *is* $fx \approx gx \approx gy \ \& \ fy \approx 0$.

(9) $\mathcal{E}(\mathbf{R}_5) = \{\varepsilon_{\mathbf{R}_5, gy \equiv 0}, \varepsilon_{\mathbf{R}_5, gy \equiv y}\}$.

(10) $\kappa(\mathbf{R}_5) = \lambda(\mathbf{R}_5) = \{\widehat{\mathcal{U}}(\mathbf{R}_5, gy \equiv 0), \widehat{\mathcal{U}}(\mathbf{R}_5, gy \equiv y)\}$.

(11) $\widehat{\mathcal{U}}(\mathbf{R}_5, gy \equiv 0) = \langle fx \approx gx \to fx \approx 0 \rangle$,

$\langle \mathbf{T}_5 \rangle \vee \langle \mathbf{R}_2 \rangle \vee \langle \mathbf{R}_4 \rangle \leq \widehat{\mathcal{U}}(\mathbf{R}_6, gy \equiv 0)$ *but* $\langle \mathbf{T}_4 \rangle \not\leq \widehat{\mathcal{U}}(\mathbf{R}_5, gy \equiv 0)$,

$\langle \mathbf{R}_2 \rangle \vee \langle \mathbf{R}_3 \rangle \vee \langle \mathbf{R}_4 \rangle \vee \langle \mathbf{R}_6 \rangle \leq \widehat{\mathcal{U}}(\mathbf{R}_5, gy \equiv y)$ *but* $\langle \mathbf{T}_5 \rangle$, $\langle \mathbf{R}_5 \rangle \not\leq \widehat{\mathcal{U}}(\mathbf{R}_5, gy \equiv y)$.

(12) $\mathbf{S} \in \widehat{\mathcal{U}}(\mathbf{R}_5, gy \equiv 0)$ *if and only if* $\mathbf{S}$ *omits* $\mathbf{T}_4$,

$\mathbf{S} \in \widehat{\mathcal{U}}(\mathbf{R}_5, gy \equiv y)$ *if and only if* $\mathbf{S}$ *omits* $\mathbf{T}_5$ *and* $\mathbf{R}_5$.

PROOF. The quasivariety $\langle \mathbf{R}_5 \rangle$ is not finitely based by Theorem 5.2. This time note that $\mathbf{R}_7 \leq \mathbf{R}_5 \times \mathbf{T}_1 \leq \mathbf{R}_5^2$, while $\mathbf{T}_4 \leq \mathbf{R}_5$. To see that $\widehat{\mathcal{U}}(\mathbf{R}_5, gy \equiv 0) = \langle fx \approx gx \to fx \approx 0 \rangle$ use the substitution $y \mapsto fx$. $\square$

Next, $\mathbf{R}_6$ is fairly easy.

PROPOSITION 5.20. *The algebra* $\mathbf{R}_6$ *satisfies the following:*

(1) $\langle \mathbf{R}_6 \rangle = \widehat{\mathcal{U}}(\mathbf{R}_1, x \equiv y) = \langle fx \approx fy \ \& \ gx \approx gy \to x \approx y \rangle$.

(2) $\langle \mathbf{T}_3 \rangle \vee \langle \mathbf{T}_4 \rangle \leq \langle \mathbf{R}_6 \rangle$.

(3) $\mathcal{Q} \leq \langle \mathbf{R}_6 \rangle$ *if and only if* $\mathcal{Q}$ *omits* $\mathbf{T}_2$, $\mathbf{T}_5$, *and* $\mathbf{R}_1$.

(4) $\mathbf{S} \in \langle \mathbf{R}_6 \rangle$ *if and only if* $\mathbf{S}$ *omits* $\mathbf{T}_2$, $\mathbf{T}_5$, $\mathbf{R}_1$, $\mathbf{R}_2$, *and* $\mathbf{R}_3$.

(5) $\langle \mathbf{R}_6 \rangle_* = \langle \mathbf{R}_6 \rangle \wedge \widehat{\mathcal{U}}(\mathbf{R}_6, fx \equiv 0) \geq \langle \mathbf{T}_3 \rangle \vee \langle \mathbf{T}_4 \rangle$.

(6) *The reflection congruences of* $\mathbf{R}_6$ *are* $\Delta, \nabla, fx \equiv 0$.

(7) $\gamma^{\mathbf{R}_6}(fx \equiv 0) = \rho^{\mathbf{R}_6}_{\langle fx \approx 0 \rangle}$ *is* $fx \equiv 0$.

(8) $\mathrm{diag}(\mathbf{R}_6)$ *is* $fx \approx gx \approx fy \ \& \ gy \approx 0$.

(9) $\mathcal{E}(\mathbf{R}_6) = \{\varepsilon_{\mathbf{R}_6, fx \equiv 0}\}$.

(10) $\kappa(\mathbf{R}_6) = \lambda(\mathbf{R}_6) = \{\widehat{\mathcal{U}}(\mathbf{R}_6, fx \equiv 0)\}$.

(11) $\langle \mathbf{T}_5 \rangle \vee \langle \mathbf{R}_2 \rangle \vee \langle \mathbf{R}_3 \rangle \vee \langle \mathbf{R}_4 \rangle \vee \langle \mathbf{R}_5 \rangle \leq \widehat{\mathcal{U}}(\mathbf{R}_6, fx \equiv 0)$ *but* $\langle \mathbf{R}_6 \rangle \not\leq \widehat{\mathcal{U}}(\mathbf{R}_6, fx \equiv 0)$.

(12) $\mathbf{S} \in \widehat{\mathcal{U}}(\mathbf{R}_6, fx \equiv 0)$ *if and only if* $\mathbf{S}$ *omits* $\mathbf{R}_6$.

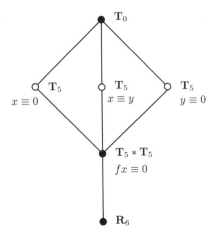

FIGURE 5.11. The congruence lattice of $\mathbf{R}_6$ where $\mathbf{R}_6 = \mathcal{F}_2/((fx \equiv gx \equiv fy) \vee (gy \equiv 0))$.

PROOF. The congruence lattice of $\mathbf{R}_6$ is in Figure 5.11. Since $\mathbf{R}_6$ is subdirectly irreducible, then $\langle \mathbf{R}_6 \rangle$ is join prime. Note that $\mathbf{T}_3, \mathbf{T}_4 \leq \mathbf{R}_6$.

Now $\mathbf{R}_6$ is a unary algebra with 0 and 2-element range, such that $\mathrm{Rows}(\mathbf{R}_6)$ is an order ideal of $\mathbf{2}^k$ and uniquely witnessed. By Theorem 5.2, $\langle \mathbf{R}_6 \rangle$ has a basis of quasi-identities in at most 2 variables. Thus $\langle \mathbf{R}_6 \rangle$ is characterized by the 2-generated quasicritical algebras it omits. Looking at Figure 5.5, we see that $\mathcal{Q} \not\leq \langle \mathbf{R}_6 \rangle$ if and only if $\mathbf{T}_2, \mathbf{T}_5$, and $\mathbf{R}_1$ are not in $\mathcal{Q}$. But this is exactly the characterization of $\widehat{\mathcal{U}}(\mathbf{R}_1, x \equiv y)$! We conclude that $\langle \mathbf{R}_6 \rangle = \widehat{\mathcal{U}}(\mathbf{R}_1, x \equiv y)$. This is the quasivariety of Corollary 5.14, and we note that it is doubly irreducible, like $\langle \mathbf{T}_5 \rangle$. ☐

The case of $\mathbf{R}_7$ is similar to that of $\mathbf{R}_1$.

PROPOSITION 5.21. *The algebra* $\mathbf{R}_7$ *satisfies the following:*

(1) $\langle \mathbf{R}_7 \rangle$ *satisfies the quasi-equations*

$$(\varepsilon_{\mathbf{T}_5, x \equiv 0}) \qquad fx \approx gx \approx 0 \rightarrow x \approx 0,$$

$$(\varepsilon_{\mathbf{R}_8, fy \equiv 0}) \qquad fx \approx gy \rightarrow fy \approx 0.$$

(2) $\langle \mathbf{T}_2 \rangle \vee \langle \mathbf{R}_1 \rangle \leq \langle \mathbf{R}_7 \rangle$.

(3) *If* $\mathcal{Q} \leq \langle \mathbf{R}_7 \rangle$, *then* $\mathcal{Q}$ *omits* $\mathbf{T}_5$, *and* $\mathbf{R}_8$.

(4) *If* $\mathbf{S} \in \langle \mathbf{R}_7 \rangle$, *then* $\mathbf{S}$ *omits* $\mathbf{T}_3, \mathbf{T}_4, \mathbf{T}_5, \mathbf{R}_8$, *and* $\mathbf{P}_2$.

(5) $\langle \mathbf{R}_7 \rangle_* = \langle \mathbf{R}_7 \rangle \wedge \widehat{\mathcal{U}}(\mathbf{R}_7, fx \equiv y) \geq \langle \mathbf{T}_2 \rangle \vee \langle \mathbf{R}_1 \rangle$.

(6) *The reflection congruences of* $\mathbf{R}_7$ *are* $\Delta, \nabla, fx \equiv 0, y \equiv fx,$
    $(fx \equiv 0) \vee (y \equiv fx), (gx \equiv x) \vee (y \equiv fx), (gx \equiv 0) \vee (fx \equiv gx)$.

(7) $\gamma^{\mathbf{R}_7}(fx \equiv 0)$ *is* $fx \equiv 0,$
    $\gamma^{\mathbf{R}_7}(y \equiv fx)$ *is* $y \equiv fx,$
    $\gamma^{\mathbf{R}_7}(fx \equiv gx) = \gamma^{\mathbf{R}_7}(gx \equiv 0)$ *is* $(fx \equiv gx) \vee (gx \equiv 0)$.

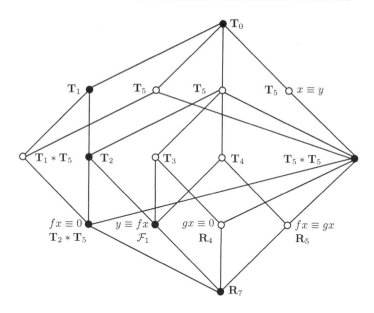

FIGURE 5.12. Con $\mathbf{R}_7$ where $\mathbf{R}_7 = \mathcal{F}_2/((fx \equiv gy) \vee (fy \equiv 0))$.

(8) $\mathrm{diag}(\mathbf{R}_7)$ *is* $fx \approx gy$ & $fy \approx 0$.

(9) $\mathcal{E}(\mathbf{R}_7) = \{\varepsilon_{\mathbf{R}_7, fx \equiv 0}, \varepsilon_{\mathbf{R}_7, fx \equiv y}, \varepsilon_{\mathbf{R}_7, gx \equiv 0}, \varepsilon_{\mathbf{R}_7, fx \equiv gx}\}$.

(10) $\kappa(\mathbf{R}_7) = \{\widehat{\mathcal{U}}(\mathbf{R}_7, fx \equiv 0), \widehat{\mathcal{U}}(\mathbf{R}_7, fx \equiv y)\}$.

(11) $\mathcal{U}(\mathbf{R}_7, fx \equiv gx) = \mathcal{U}(\mathbf{R}_7, gx \equiv 0) = \langle gx \approx 0 \rangle = \langle \mathbf{T}_5 \rangle$,
$\widehat{\mathcal{U}}(\mathbf{R}_7, fx \equiv 0) = \langle fx \approx 0 \rangle = \langle \mathbf{T}_1 \rangle \vee \langle \mathbf{T}_5 \rangle$,
$\langle \mathbf{R}_2 \rangle \vee \langle \mathbf{R}_3 \rangle \vee \langle \mathbf{R}_6 \rangle \leq \widehat{\mathcal{U}}(\mathbf{R}_7, fx \equiv 0)$
but $\langle \mathbf{T}_5 \rangle, \langle \mathbf{R}_7 \rangle \not\leq \widehat{\mathcal{U}}(\mathbf{R}_7, fx \equiv 0)$.

(12) $\mathbf{S} \in \mathcal{U}(\mathbf{R}_7, gx \equiv 0)$ *if and only if* $\mathbf{S}$ *omits* $\mathbf{T}_1$,
$\mathbf{S} \in \widehat{\mathcal{U}}(\mathbf{R}_7, fx \equiv 0)$ *if and only if* $\mathbf{S}$ *omits* $\mathbf{T}_3, \mathbf{T}_4$, *and* $\mathcal{F}_1$,
$\mathbf{S} \in \widehat{\mathcal{U}}(\mathbf{R}_7, fx \equiv y)$ *if and only if* $\mathbf{S}$ *omits* $\mathbf{T}_5, \mathbf{R}_4, \mathbf{R}_5$, *and* $\mathbf{R}_7$.

PROOF. The congruence lattice of $\mathbf{R}_7$ is in Figure 5.12. We do not know whether the quasivariety $\langle \mathbf{R}_7 \rangle$ is finitely based. The inclusions $\langle \mathbf{R}_1 \rangle \leq \langle \mathbf{R}_7 \rangle \leq \langle \mathbf{R}_4 \rangle \wedge \langle \mathbf{R}_5 \rangle$ and the dependency $\langle \mathbf{R}_7 \rangle \leq \langle \mathbf{T}_5 \rangle \vee \langle \mathcal{F}_1 \rangle$ are explained in Table 5.3.

In the variety $\mathcal{M}$ the quasi-equations $\varepsilon_{\mathbf{R}_7, fx \equiv 0}$, $\varepsilon_{\mathbf{R}_1, gx \equiv 0}$ and $\varepsilon_{\mathbf{R}_1, fx \equiv gx}$ are equivalent to the equations that are their conclusions, i.e., $fx \approx 0$, $gx \approx 0$ and $fx \approx gx$, respectively. To see this, use the substitution $y \mapsto fx$ and the equations of $\mathcal{M}$. □

We conclude this section with $\mathbf{R}_8$.

PROPOSITION 5.22. *The algebra* $\mathbf{R}_8$ *satisfies the following:*

(1) $\langle \mathbf{R}_8 \rangle$ *satisfies the quasi-equations*

$$(\varepsilon_{\mathbf{P}_2, fx \equiv 0}) \qquad fx \approx gy \ \& \ gx \approx fy \to fx \approx 0,$$

$$(\varepsilon_{\mathbf{R}_1, x \equiv y}) \qquad fx \approx fy \ \& \ gx \approx gy \to x \approx y,$$

and more generally the quasi-equations $\varepsilon_{\mathbf{P}_n, fx_1 \equiv gx_1}$ *for* $n \geq 2$ *of Theorem 5.26.*

(2) *The quasicritical algebras in* $\langle \mathbf{R}_8 \rangle$ *are* $\mathbf{T}_1$, $\mathcal{F}_1$, *and* $\mathbf{R}_8$.

(3) *If* $\mathcal{Q} \leq \langle \mathbf{R}_8 \rangle$, *then* $\mathcal{Q}$ *omits* $\mathbf{T}_2$, $\mathbf{T}_4$, $\mathbf{T}_5$, $\mathbf{R}_1$, *and* $\mathbf{P}_2$.

(4) *If* $\mathbf{S} \in \langle \mathbf{R}_8 \rangle$, *then* $\mathbf{S}$ *omits* $\mathbf{T}_2$, $\mathbf{T}_3$, $\mathbf{T}_4$, $\mathbf{T}_5$, $\mathbf{R}_1$, *and* $\mathbf{P}_2$.

(5) $\langle \mathbf{R}_8 \rangle_* = \langle \mathbf{R}_8 \rangle \wedge \widehat{\mathcal{U}}(\mathbf{R}_8, fy \equiv 0) = \langle \mathcal{F}_1 \rangle$.

(6) *The reflection congruences of* $\mathbf{R}_8$ *are* Δ, ∇, $fy \equiv 0$, $fx \equiv y$, $fx \equiv fy \equiv 0$, $y \equiv fx \equiv fy \equiv 0$, $gx \equiv fx \equiv fy \equiv 0$, $(y \equiv fx \equiv fy \equiv 0) \vee (gx \equiv x)$.

(7) $\gamma^{\mathbf{R}_8}(fy \equiv 0) = \rho^{\mathbf{R}_8}_{\langle fy \equiv 0 \rangle}$ *is* $fy \equiv 0$,
$\gamma^{\mathbf{R}_8}(fx \equiv fy) = \gamma^{\mathbf{R}_8}(fx \equiv 0)$ *is* $fx \equiv fy \equiv 0$,
$\gamma^{\mathbf{R}_8}(fx \equiv gx) = \gamma^{\mathbf{R}_8}(gx \equiv fy) = \gamma^{\mathbf{R}_8}(gx \equiv 0)$ *is* $gx \equiv fx \equiv fy \equiv 0$.

(8) $\mathrm{diag}(\mathbf{R}_8)$ *is* $fx \approx gy$.

(9) $\mathcal{E}(\mathbf{R}_8) = \{\varepsilon_{\mathbf{R}_8, fy \equiv 0}, \varepsilon_{\mathbf{R}_8, fx \equiv fy}, \varepsilon_{\mathbf{R}_8, fx \equiv 0}, \varepsilon_{\mathbf{R}_8, fx \equiv gx}, \varepsilon_{\mathbf{R}_8, gx \equiv fy}, \varepsilon_{\mathbf{R}_8, gx \equiv 0}\}$.

(10) $\kappa(\mathbf{R}_8) = \{\widehat{\mathcal{U}}(\mathbf{R}_8, fy \equiv 0)\}$.

(11) $\mathcal{U}(\mathbf{R}_8, fx \equiv gx) = \mathcal{U}(\mathbf{R}_8, gx \equiv fy) = \mathcal{U}(\mathbf{R}_8, gx \equiv 0) = \langle \mathbf{T}_5 \rangle$,
$\mathcal{U}(\mathbf{R}_8, fx \equiv fy) = \mathcal{U}(\mathbf{R}_8, fx \equiv 0)$,
$\mathcal{U}(\mathbf{R}_8, fx \equiv gx) < \mathcal{U}(\mathbf{R}_8, fx \equiv fy) < \widehat{\mathcal{U}}(\mathbf{R}_8, fy \equiv 0)$,
$\langle \mathbf{T}_5 \rangle \vee \langle \mathbf{R}_7 \rangle \leq \widehat{\mathcal{U}}(\mathbf{R}_8, fy \equiv 0)$ *but* $\langle \mathbf{R}_8 \rangle \not\leq \widehat{\mathcal{U}}(\mathbf{R}_8, fy \equiv 0)$.

(12) $\mathbf{S} \in \mathcal{U}(\mathbf{R}_8, fx \equiv gx)$ *if and only if* $\mathbf{S}$ *omits* $\mathbf{T}_1$,
$\mathbf{S} \in \mathcal{U}(\mathbf{R}_8, fx \equiv 0)$ *if and only if* $\mathbf{S}$ *omits* $\mathbf{T}_3$, $\mathbf{T}_4$, *and* $\mathcal{F}_1$,
$\mathbf{S} \in \widehat{\mathcal{U}}(\mathbf{R}_8, fy \equiv 0)$ *if and only if* $\mathbf{S}$ *omits* $\mathbf{T}_3$, $\mathbf{T}_4$, $\mathbf{R}_8$, *and* $\mathbf{P}_2$.

PROOF. The congruence lattice of $\mathbf{R}_8$, appears in Figure 5.13. Note that the quasi-equation $\varepsilon_{\mathbf{P}_2, fx \equiv 0}$ given in (1) is equivalent to the conjunction of $\varepsilon_{\mathbf{T}_4, fx \equiv 0}$ and $\varepsilon_{\mathbf{P}_2, fx \equiv gx}$. We don't know whether the quasivariety $\langle \mathbf{R}_8 \rangle$ is finitely based.

For (5) apply Lemma 5.11. We have $\langle \mathcal{F}_1 \rangle \leq \langle \mathbf{R}_8 \rangle \leq \widehat{\mathcal{U}}(\mathbf{R}_1, x \equiv y)$ and $\langle \mathcal{F}_1 \rangle \leq \widehat{\mathcal{U}}(\mathbf{R}_8, fy \equiv 0)$, whence $\langle \mathcal{F}_1 \rangle = \langle \mathbf{R}_8 \rangle \wedge \widehat{\mathcal{U}}(\mathbf{R}_8, fy \equiv 0)$. Claim (2) follows from this. $\qquad \square$

5.4. Uncountably Many Subquasivarieties

In this section we show that there are uncountably many subquasivarieties contained in $\langle \mathbf{T}_3 \rangle$. We begin by identifying an infinite class of quasicritical algebras in $\mathcal{M}$.

To see that there are infinitely many quasicritical algebras set $\mathcal{F}_n$ to be the free algebra generated by $\{x_1, x_2, \ldots, x_n\}$. Consider, for $n \geq 2$, the

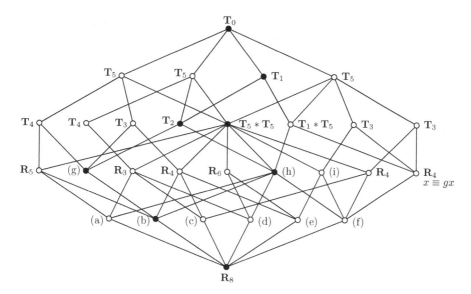

FIGURE 5.13. Con $\mathbf{R}_8$ where $\mathbf{R}_8 = \mathcal{F}_2/(fx \equiv gy)$. The labels for the diagram are: (a) $fx \equiv gx$ with the non-quasicritical quotient $\mathbf{R}_8/(a) \leq \mathbf{T}_4^2$, (b) $fy \equiv 0$ with $\mathbf{R}_8/(b) \cong \mathbf{R}_7$, (c) $gx \equiv fy$ with $\mathbf{R}_8/(c) \cong \mathbf{P}_2$, (d) $fx \equiv fy$ with the non-quasicritical quotient $\mathbf{R}_8/(d) \leq \mathbf{T}_1 \times \mathbf{T}_4^2$, (e) $gx \equiv 0$ with the non-quasicritical quotient $\mathbf{R}_8/(e) \leq \mathbf{T}_3^2$, (f) $fx \equiv 0$ with the non-quasicritical quotient $\mathbf{R}_8/(f) \cong \mathbf{T}_2 * \mathbf{T}_3$, (g) $fx \equiv y$ with $\mathbf{R}_8/(g) \cong \mathcal{F}_1$, (h) join reducible congruence with non-quasicritical quotient $\mathbf{R}_8/(h) \cong \mathbf{T}_2 * \mathbf{T}_5$. (i) join reducible congruence with non-quasicritical quotient $\mathbf{R}_8/(i) \cong \mathbf{T}_3 * \mathbf{T}_5$.

algebra $\mathbf{P}_n$ that is the homomorphic image of $\mathcal{F}_n$ determined by the join of the congruences $fx_1 \equiv gx_n$ and $fx_{i+1} \equiv gx_i$ for $1 \leq i < n$. When drawn, $\mathbf{P}_n$ looks like an n-pointed star; see Figures 5.14 and 5.18. The next lemma shows that each $\mathbf{P}_n$ $(n \geq 2)$ is quasicritical. Corollary 5.27 uses the algebras $\mathbf{P}_n$ to show that $L_q(\mathcal{M})$ is uncountable.

LEMMA 5.23. *There are infinitely many quasicritical algebras in* $\mathcal{M}$.

PROOF. Fix $n \geq 2$. If $\mathbf{P}_n$ is not quasicritical, then it is a subdirect product of proper subalgebras. Every maximal subalgebra of $\mathbf{P}_n$ is isomorphic to the algebra $\mathbf{A}$ with universe $P_n \setminus \{x_1\}$. That is,

$$A = \{x_2, x_3, \ldots, x_n\} \cup \{gx_1, gx_2, \ldots, gx_n\} \cup \{0\}.$$

Assume that $b_1, b_2, \ldots, b_n$ in A satisfy $fb_1 = gb_n$ and $fb_{i+1} = gb_i$ for $1 \leq i < n$. Set $B = \{b_1, b_2, \ldots, b_n\}$. We show that $B = \{0\}$. To see this, first assume for some $i < n$ that $x_i \in B$, say, $x_i = b_j$. Then $fb_{j+1} = gb_j = gx_i$.

But the only element in P_n that satisfies $fy = gx_i$ is x_{i+1}. Thus b_{j+1} must be x_{i+1}. Continuing in this manner, we obtain $x_n = b_\ell$ for some ℓ. However, if $x_n = b_\ell$, then $fb_u = gb_\ell = gx_n$ with $u \in \{1, 1+\ell\}$, while no y in A satisfies $fy = gx_n$. Thus x_n is not in B. This means no x_i is in B. Now assume that some gx_i is in B. If y satisfies $fy = gx_i$, then y is some x_j. Since no x_j is in B, neither is gx_i. Thus $B = \{0\}$.

To see that $\mathbf{P}_n$ is quasicritical, assume that $h\colon \mathbf{P}_n \to \mathbf{A}^s$ is an embedding. Set $h(x_i) = \langle a_1^i, a_2^i, \ldots, a_s^i \rangle$. Because we have $fx_1 = gx_n$ and $fx_{i+1} = gx_i$ for $1 \le i < n$, applying h to these elements gives $fa_j^1 = ga_j^n$ and $fa_j^{i+1} = g_j^i$ for $1 \le i < n$. By the above argument each a_j^i is 0, and so h is not an embedding.   □

Figure 5.14 illustrates $\mathbf{P}_3$ and two homomorphic images of it that are quasicritical. The verification that they are quasicritical was done with the Universal Algebra Calculator.

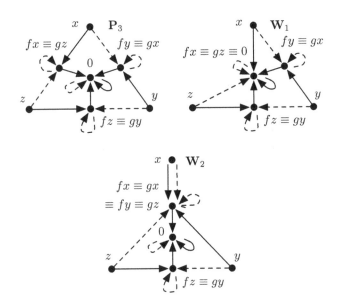

FIGURE 5.14. Some quasicritical homomorphic images of $\mathcal{F}_3$

The next lemma includes the fact that the algebras $\mathbf{P}_n$ are all in $\langle \mathbf{T}_3 \rangle$.

PROPOSITION 5.24. *The algebra* $\mathbf{T}_3$ *satisfies the following:*

(1) $\langle \mathbf{T}_3 \rangle = \langle fx \approx gx \to fx \approx 0,\ fx \approx fy\ \&\ gx \approx gy \to x \approx y \rangle$.

(2) $\{\mathbf{R}_8, \mathcal{F}_1, \mathbf{T}_1\} \cup \{\mathbf{P}_n : n \ge 2\} \subseteq \langle \mathbf{T}_3 \rangle$.

(3) $\mathfrak{Q} \le \langle \mathbf{T}_3 \rangle$ *if and only if* $\mathfrak{Q}$ *omits* $\mathbf{T}_2$, $\mathbf{T}_4$, $\mathbf{T}_5$, *and* $\mathbf{R}_1$.

(4) $\mathbf{S} \in \langle \mathbf{T}_3 \rangle$ *if and only if* $\mathbf{S}$ *omits* $\mathbf{T}_2$, $\mathbf{T}_4$, $\mathbf{T}_5$, $\mathbf{R}_1$, *and* $\mathbf{R}_2$.

(5) $\langle \mathbf{T}_3 \rangle_* = \langle \mathbf{T}_3 \rangle \wedge \widehat{\mathcal{U}}(\mathbf{T}_3, fx \equiv 0) \ge \bigvee_{n \ge 2} \langle \mathbf{P}_n \rangle$.

(6) *All three congruences of* $\mathbf{T}_3$ *are reflection congruences:* Δ, ∇, $fx \equiv 0$.

(7) $\gamma^{\mathbf{T}_3}(fx \equiv 0) = \rho^{\mathbf{T}_3}_{\langle fx \approx 0 \rangle}$ *is* $fx \equiv 0$.

(8) $\mathrm{diag}(\mathbf{T}_3)$ *is* $gx \approx 0$.

(9) $\mathcal{E}(\mathbf{T}_3) = \{\varepsilon_{\mathbf{T}_3, fx \equiv 0}\}$.

(10) $\kappa(\mathbf{T}_3) = \lambda(\mathbf{T}_3) = \{\widehat{\mathcal{U}}(\mathbf{T}_3, fx \equiv 0)\}$.

(11) *The quasicritical algebras in* $\widehat{\mathcal{U}}(\mathbf{T}_3, fx \equiv 0)$ *are* $\mathbf{T}_5$ *and every quasicritical algebra that satisfies* $gx \approx 0 \rightarrow x \approx 0$, *so that* $\widehat{\mathcal{U}}(\mathbf{T}_3, fx \equiv 0) = \langle \mathbf{T}_5 \rangle \vee \langle gx \approx 0 \rightarrow x \approx 0 \rangle$.

(12) $\mathbf{S} \in \widehat{\mathcal{U}}(\mathbf{T}_3, fx \equiv 0)$ *if and only if* $\mathbf{S}$ *omits* $\mathbf{T}_3$.

(13) *See* Figure 5.15 *for the segment of* $\mathrm{L_q}(\mathcal{M})$ *related to* $\mathbf{T}_3$.

PROOF. The congruence lattice of $\mathbf{T}_3$ is a 3-element chain with its atom defined by $fx \equiv 0$ and $\mathbf{T}_3/(fx \equiv 0) \cong \mathbf{T}_5$.

As observed earlier, by Theorem 5.2 we know that $\langle \mathbf{T}_3 \rangle$ has a 2-variable basis of quasi-identities, and thus is determined by the 2-generated quasicritical algebras it omits (Theorem 3.10). These are $\mathbf{T}_2$, $\mathbf{T}_4$, $\mathbf{T}_5$, and $\mathbf{R}_1$. Once we know that, we obtain the quasi-equational basis consisting of $\varepsilon_{\mathbf{T}_4, fx \equiv 0}$ and $\varepsilon_{\mathbf{R}_1, x \equiv y}$ given in (1).

To see that $\mathbf{P}_n \in \langle \mathbf{T}_3 \rangle$ for $n \geq 2$ let c_i in $\mathbf{T}_3{}^n$ be defined for $1 \leq i \leq n-1$ by

$$c_i(j) = \begin{cases} x & \text{if } j = i, \\ fx & \text{if } j = i+1, \\ 0 & \text{otherwise,} \end{cases}$$

and let

$$c_n(j) = \begin{cases} x & \text{if } j = n, \\ fx & \text{if } j = 1, \\ 0 & \text{otherwise.} \end{cases}$$

Then $f(c_1) = g(c_n)$ and $f(c_i) = g(c_{i-1})$ otherwise. Thus the subalgebra generated by $\{c_1, \ldots, c_n\}$ is $\mathbf{P}_n$.

Let us show that (11) holds. Since $\widehat{\mathcal{U}}(\mathbf{T}_3, fx \equiv 0) = \langle gx \approx 0 \rightarrow fx \approx 0 \rangle$, it follows that $\langle \mathbf{T}_5 \rangle \vee \langle gx \approx 0 \rightarrow x \approx 0 \rangle \leq \widehat{\mathcal{U}}(\mathbf{T}_3, fx \equiv 0)$. For the reverse inclusion, assume that $\mathbf{A}$ is a quasicritical algebra in $\widehat{\mathcal{U}}(\mathbf{T}_3, fx \equiv 0)$. If there exists $a \neq 0$ in A with $g(a) = 0$, then also $f(a) = 0$ by $\varepsilon_{\mathbf{T}_3, fx \equiv 0}$, so that $\mathrm{Sg}(a) \cong \mathbf{T}_5$. However, $a \notin \mathrm{Sg}(A \setminus \{a\})$. For if $f(b) = a$, then using the identities of $\mathcal{M}$ we have $a = f(b) = gf(b) = g(a) = 0$, a contradiction. Likewise, if $g(b) = a$, then $a = g(b) = gg(b) = g(a) = 0$. Thus $\mathbf{A} \cong \mathbf{T}_5 * (\mathbf{A} \setminus \{a\})$, and by its quasicriticality $\mathbf{A} \cong \mathbf{T}_5$. On the other hand, if $g(a) \neq 0$ for all nonzero $a \in A$, then $\mathbf{A}$ satisfies $gx \approx 0 \rightarrow x \approx 0$. Thus every quasicritical algebra in $\widehat{\mathcal{U}}(\mathbf{T}_3, fx \equiv 0)$ is in either $\langle \mathbf{T}_5 \rangle$ or $\langle gx \approx 0 \rightarrow x \approx 0 \rangle$, and we conclude that equality holds for the join.

It is easy to see that $gx \approx 0 \rightarrow x \approx 0$ is equivalent to the conjunction of the quasi-equations $\varepsilon_{\mathbf{T}_3, fx \equiv 0}$, which is $gx \approx 0 \rightarrow fx \approx 0$, and $\varepsilon_{\mathbf{T}_5, x \equiv 0}$, which is $fx \approx gx \approx 0 \to x \approx 0$. Thus we have $\langle gx \approx 0 \to x \approx 0 \rangle = \widehat{\mathcal{U}}(T_3, fx \equiv 0) \wedge \widehat{\mathcal{U}}(\mathbf{T}_5, x \equiv 0)$, and it follows that the lower cover $\langle \mathbf{T}_3 \rangle_* = \langle \mathbf{T}_3 \rangle \wedge \langle gx \approx 0 \to x \approx 0 \rangle$. These facts are reflected in Figure 5.15. □

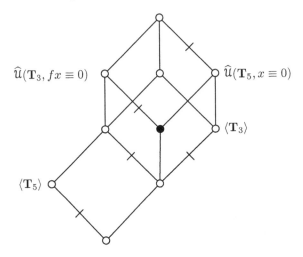

FIGURE 5.15. The segment of $L_q(\mathcal{M})$ related to $\langle \mathbf{T}_3 \rangle$. The solid dot marks the quasivariety $\langle gx \approx 0 \to x \approx 0 \rangle$. Ticks indicate covers.

The congruence lattice of our next algebra, $\mathbf{P}_2$, is illustrated in Figure 5.16.

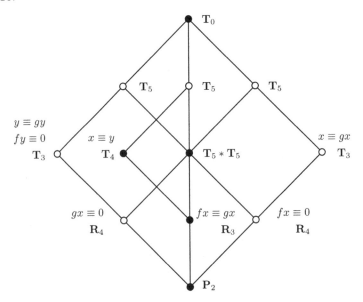

FIGURE 5.16.   Con $\mathbf{P}_2$ where $\mathbf{P}_2 = \mathcal{F}_2/((fx \equiv gy) \vee (gx \equiv fy))$.

PROPOSITION 5.25. *The algebra* $\mathbf{P}_2$ *satisfies the following:*

(1) $\langle \mathbf{P}_2 \rangle$ *satisfies the quasi-equations*

$$(\varepsilon_{\mathbf{T}_3, fx \equiv 0}) \qquad gx \approx 0 \to fx \approx 0,$$

$$(\varepsilon_{\mathbf{R}_1, x \equiv y}) \qquad fx \approx fy \ \& \ gx \approx gy \to x \approx y,$$

and all the quasi-equations $\varepsilon_{\mathbf{P}_n, fx_1 \equiv gx_1}$ *for* n *odd.*

(2) $\mathbf{T}_1, \mathcal{F}_1, \mathbf{R}_8 \in \langle \mathbf{P}_2 \rangle$.

(3) *If* $\mathcal{Q} \le \langle \mathbf{P}_2 \rangle$, *then* $\mathcal{Q}$ *omits* $\mathbf{T}_2, \mathbf{T}_3, \mathbf{T}_4, \mathbf{T}_5$, *and* $\mathbf{R}_1$.

(4) *If* $\mathbf{S} \in \langle \mathbf{P}_2 \rangle$, *then* $\mathbf{S}$ *omits* $\mathbf{T}_2, \mathbf{T}_3, \mathbf{T}_4, \mathbf{T}_5$, *and* $\mathbf{R}_1$.

(5) $\langle \mathbf{P}_2 \rangle_* = \langle \mathbf{P}_2 \rangle \wedge \widehat{\mathcal{U}}(\mathbf{P}_2, fx \equiv gx) \ge \langle \mathbf{R}_8 \rangle$.

(6) *The reflection congruences of* $\mathbf{P}_2$ *are* Δ, ∇, $fx \equiv gx$, $x \equiv y$, $fx \equiv gx \equiv 0$.

(7) $\gamma^{\mathbf{P}_2}(fx \equiv gx) = \rho^{\mathbf{P}_2}_{\langle fx \approx gx \rangle}$ *is* $fx \equiv gx$,
$\gamma^{\mathbf{P}_2}(fx \equiv 0) = \gamma^{\mathbf{P}_2}(gx \equiv 0)$ *is* $fx \equiv gx \equiv 0$.

(8) $\operatorname{diag}(\mathbf{P}_2)$ *is* $fx \approx gy \ \& \ gx \approx fy$.

(9) $\mathcal{E}(\mathbf{P}_2) = \{\varepsilon_{\mathbf{P}_2, fx \equiv 0}, \varepsilon_{\mathbf{P}_2, fx \equiv gx}, \varepsilon_{\mathbf{P}_2, gx \equiv 0}\}$.

(10) $\kappa(\mathbf{P}_2) = \{\widehat{\mathcal{U}}(\mathbf{P}_2, fx \equiv gx)\}$.

(11) $\mathcal{U}(\mathbf{P}_2, fx \equiv 0) = \mathcal{U}(\mathbf{P}_2, gx \equiv 0)$ *and this quasivariety is meet reducible:* $\mathcal{U}(\mathbf{P}_2, fx \equiv 0) = \widehat{\mathcal{U}}(\mathbf{T}_4, fx \equiv 0) \wedge \widehat{\mathcal{U}}(\mathbf{P}_2, fx \equiv gx)$.
$\langle \mathbf{T}_5 \rangle \vee \langle \mathbf{R}_7 \rangle \vee \langle \mathbf{R}_8 \rangle \vee \langle \mathbf{P}_3 \rangle \ \le \ \mathcal{U}(\mathbf{P}_2, fx \equiv 0)$ *but* $\langle \mathbf{T}_4 \rangle, \langle \mathbf{P}_2 \rangle \nleq \mathcal{U}(\mathbf{P}_2, fx \equiv 0)$,
$\langle \mathbf{T}_5 \rangle \vee \langle \mathbf{R}_3 \rangle \vee \langle \mathbf{R}_5 \rangle \vee \langle \mathbf{P}_3 \rangle \ \le \ \widehat{\mathcal{U}}(\mathbf{P}_2, fx \equiv gx)$ *but* $\langle \mathbf{P}_2 \rangle \nleq \widehat{\mathcal{U}}(\mathbf{P}_2, fx \equiv gx)$.

(12) $\mathbf{S} \in \mathcal{U}(\mathbf{P}_2, fx \equiv 0)$ *if and only if* $\mathbf{S}$ *omits* $\mathbf{T}_3, \mathbf{T}_4$, *and* $\mathbf{P}_2$,
$\mathbf{S} \in \widehat{\mathcal{U}}(\mathbf{P}_2, fx \equiv gx)$ *if and only if* $\mathbf{S}$ *omits* $\mathbf{T}_3$ *and* $\mathbf{P}_2$.

(13) *See* Figure 5.17 *for the segment of* $L_q(\mathcal{M})$ *related to* $\mathbf{P}_2$.

PROOF. The fact that $\mathbf{P}_2$ satisfies $\varepsilon_{\mathbf{P}_n, fx_1 \equiv gx_1}$ for n odd follows from Theorem 5.26(3). We do not know whether the quasivariety $\langle \mathbf{P}_2 \rangle$ is finitely based.

To see that $\mathcal{U}(\mathbf{P}_2, fx \equiv 0) = \widehat{\mathcal{U}}(\mathbf{T}_4, fx \equiv 0) \wedge \widehat{\mathcal{U}}(\mathbf{P}_2, fx \equiv gx)$, note that $\widehat{\mathcal{U}}(\mathbf{T}_4, fx \equiv 0)$ is characterized by omitting $\mathbf{T}_4$, $\widehat{\mathcal{U}}(\mathbf{P}_2, fx \equiv gx)$ by omitting $\mathbf{T}_3$ and $\mathbf{P}_2$, while $\mathcal{U}(\mathbf{P}_2, fx \equiv 0)$ is characterized by omitting all three. Alternatively, straightforward arguments show that $\varepsilon_{\mathbf{P}_2, fx \equiv 0}$ is equivalent to the conjunction of $\varepsilon_{\mathbf{T}_4, fx \equiv 0}$ and $\varepsilon_{\mathbf{P}_2, fx \equiv gx}$. $\square$

Note that the meet reducible quasivariety $\langle \varepsilon_{\mathbf{P}_2, fx \equiv 0} \rangle$ in part (11) provides our first example that the converse of Theorem 2.30 is false, that a quasivariety $\langle \varepsilon_{\mathbf{T}, \alpha} \rangle$ with $\mathbf{T}$ quasicritical and $\alpha \succ \Delta$ need not be meet irreducible when it is not in $\kappa(\mathbf{T})$. This example is generalized in Proposition 5.29, and more examples occur in Proposition 7.9.

For $n \geq 2$ let ϕ_n denote the quasi-equation $\varepsilon_{\mathbf{P}_n, fx_1 \equiv gx_1}$, which is

$$fx_2 \approx gx_1 \,\&\, fx_3 \approx gx_2 \,\&\, \dots \,\&\, fx_n \approx gx_{n-1} \,\& fx_1 \approx gx_n$$
$$\to fx_1 \approx gx_1.$$

THEOREM 5.26. *The algebras $\mathbf{P}_n$ with $n \geq 2$ satisfy the following:*

(1) $\mathbf{P}_n \leq \mathbf{T}_3{}^n$ *as a subdirect product.*
(2) $\mathbf{P}_n \leq \mathbf{T}_3^{n-1} \times \mathbf{T}_4$ *as a subdirect product.*
(3) $\mathbf{P}_n$ *satisfies $\varepsilon_{\mathbf{T}_4, fx \equiv 0}$, and hence $\mathbf{P}_n \notin \langle \mathbf{T}_4 \rangle$.*
(4) $\mathbf{P}_n$ *satisfies ϕ_m if and only if $n \nmid m$.*
(5) $\langle \phi_m \rangle \leq \langle \phi_n \rangle$ *if and only if $n \mid m$.*

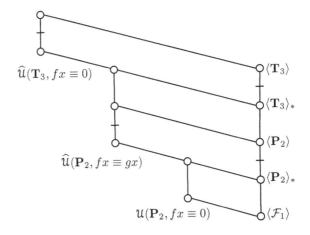

FIGURE 5.17. Segment of $L_q(\mathcal{M})$ related to $\langle \mathbf{P}_2 \rangle$

PROOF. Item (1) is shown in Proposition 5.24. The second item was shown for $\mathbf{P}_3$ by computing its congruence lattice using the Universal Algebra Calculator. The extension to $\mathbf{P}_n$ is tedious and left to the reader, the crucial observation being that $\mathbf{P}_n$ has a unique congruence θ with $\mathbf{P}_n/\theta \cong \mathbf{T}_4$. A more straightforward calculation yields (3).

To prove (4), consider various substitutions $\sigma : \{x_1, \dots, x_m\} \to \mathbf{P}_n$ such that the hypotheses of ϕ_m hold:

$$f\sigma(x_2) = g\sigma(x_1) \,\&\, f\sigma(x_3) = g\sigma(x_2) \,\&\, \cdots$$
$$\&\, f\sigma(x_m) = g\sigma(x_{m-1}) \,\&\, f\sigma(x_1) = g\sigma(x_m).$$

Let $a_1, \dots, a_n$ denote the generators of $\mathbf{P}_n$. Note that for any element $b \in \mathbf{P}_n \setminus \{a_1, \dots, a_n\}$ we have $f(b) = 0$, and that $\mathbf{P}_n$ satisfies $gx \approx 0 \to x \approx 0$. So if $\sigma(x_1) \notin \{a_1, \dots, a_n\}$, then from hypotheses of ϕ_m we get $0 = f\sigma(x_1) = g\sigma(x_m)$, which implies $\sigma(x_m) = 0$. Repeating the argument, $0 = f\sigma(x_m) = g\sigma(x_{m-1})$ whence $\sigma(x_{m-1}) = 0$. Likewise, we obtain $\sigma(x_{m-2}) = \cdots = \sigma(x_1) = 0$. In that case the conclusion of ϕ_m holds as well.

Thus we may assume that say $\sigma(x_1) = a_1$, by the circular symmetry of $\mathbf{P}_n$. But the only element $b \in \mathbf{P}_n$ with $g(b) = f(a_1)$ is $b = a_2$, whence $\sigma(x_2) = a_2$. Continuing in this fashion, assuming that the hypotheses of ϕ_m hold, we see that $\sigma(x_k) = a_{k \bmod n}$ for $1 \leq k \leq m$.

If $n \mid m$, then the hypotheses are satisfied, but the conclusion $f(x_1) \approx f(x_2)$ fails since $f(a_1) \neq f(a_2)$. Thus $\mathbf{P}_n$ fails ϕ_m when $n \mid m$. On the other hand, if $n \nmid m$, then the hypotheses are satisfied only for the trivial substitution with $\sigma(x_k) = 0$ for all k, whence $\mathbf{P}_n$ satisfies ϕ_m. Thus (4) holds.

In particular, if $n \nmid m$, then the algebra $\mathbf{P}_n$ satisfies ϕ_m but not ϕ_n, and hence $\langle \phi_m \rangle \not\leq \langle \phi_n \rangle$. This is one direction of (5). For the reverse direction, assume $n \mid m$ and let $\mathbf{B}$ be an algebra that satisfies ϕ_m. Any substitution $\sigma : \{x_1, \ldots, x_n\} \to \mathbf{B}$ can be extended cyclically to a substitution $\widehat{\sigma} : \{x_1, \ldots, x_m\} \to \mathbf{B}$ in such a way that if the hypotheses of ϕ_n are satisfied under σ, then the hypotheses of ϕ_m are satisfied under $\widehat{\sigma}$. Since the two quasi-equations have the same conclusion, we conclude that $\mathbf{B}$ satisfies ϕ_n as well.   □

COROLLARY 5.27. *The quasivariety $\langle \mathbf{T}_3 \rangle$ has $2^{\aleph_0}$ subquasivarieties.*

PROOF. For q prime, $\langle \phi_q \rangle$ omits $\mathbf{P}_n$ for $n = q$ only. When I is an arbitrary set of primes, $\bigwedge_{q \in I} \langle \phi_q \rangle$ omits $\mathbf{P}_n$ exactly when $n \in I$.   □

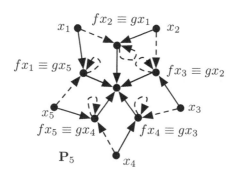

FIGURE 5.18. The algebra $\mathbf{P}_5$

An alternate proof of Corollary 5.27 is based on the following observation.

LEMMA 5.28. *If $q \geq 2$ is prime, then $\mathbf{P}_q$ has a unique minimal nonzero reflection congruence.*

PROOF. Figure 5.16 shows this for $\mathbf{P}_2$. Consider an algebra $\mathbf{P}_q$ with $q \geq 3$ prime. ($\mathbf{P}_5$ is illustrated in Figure 5.18.) The atoms of Con $\mathbf{P}_q$ are

- α_{jk} given by $fx_j \equiv fx_k$ for $j \neq k$,
- β_j given by $fx_j \equiv 0$ for all j.

Let $\alpha = \bigvee_{j \neq k} \alpha_{jk}$ and $\beta = \bigvee_j \beta_j$, and note that $\alpha < \beta$. This is illustrated for $q = 3$ in Figure 5.19.

Any characteristic congruence containing some α_{jk} must be above α (using q prime), and likewise any characteristic congruence containing some β_j is above β. Since $\alpha < \beta$, any nonzero characteristic congruence contains α. By Theorem 2.23, every reflection congruence is characteristic, and thus above α. Hence $\gamma^{\mathbf{P}_q}(\alpha)$ is the unique minimal nonzero reflection congruence of $\mathbf{P}_q$. (In fact $\gamma^{\mathbf{P}_q}(\alpha) = \alpha$ because $\mathbf{P}_q/\alpha$ satisfies ϕ_q, but that is not needed.) □

Therefore by Corollary 2.33, $\langle \mathbf{P}_q \rangle$ for $q \geq 2$ prime is completely join prime in $\mathrm{L_q}(\mathcal{M})$. Also $\langle \mathbf{P}_q \rangle$ and $\langle \mathbf{P}_r \rangle$ are incomparable for distinct primes q and r. Since these are all in $\langle \mathbf{T}_3 \rangle$, that quasivariety has $2^{\aleph_0}$ subquasivarieties.

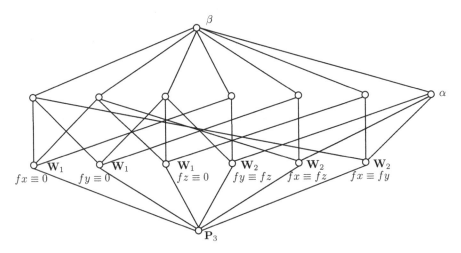

FIGURE 5.19. An ideal in the congruence lattice of $\mathbf{P}_3$ where $\mathbf{P}_3 = \mathcal{F}_3/((fx \equiv gy) \vee (fy \equiv gz) \vee (fz \equiv gx))$.

Let us pause to note that the algebras $\mathbf{P}_n$ provide more examples of meet reducible quasivarieties $\langle \varepsilon_{\mathbf{T},\alpha} \rangle$ with $\mathbf{T}$ quasicritical and $\alpha \succ \Delta$, generalizing Proposition 5.25(11).

PROPOSITION 5.29. *For $n \geq 2$,*

$$\mathcal{U}(\mathbf{P}_n, fx \equiv 0) = \widehat{\mathcal{U}}(\mathbf{P}_n, fx \equiv gx) \wedge \widehat{\mathcal{U}}(\mathbf{T}_4, fx \equiv 0).$$

PROOF. Straightforward arguments, using the circular symmetry of $\mathbf{P}_n$, show that $\varepsilon_{\mathbf{P}_n, fx \equiv 0}$ is equivalent to the conjunction of $\varepsilon_{\mathbf{P}_n, fx \equiv gx}$ and $\varepsilon_{\mathbf{T}_4, fx \equiv 0}$. That is, $\mathrm{diag}(\mathbf{P}_n) \to fx \approx 0$ is equivalent to the conjunction of $\mathrm{diag}(\mathbf{P}_n) \to fx \approx gx$ and $fx \approx gx \to fx \approx 0$. □

5.5. Q-Universality of $\mathcal{M}$

A quasivariety $\mathcal{K}$ is said to be *Q-universal* if, for every quasivariety $\mathcal{Q}$ of finite type, $\mathrm{L_q}(\mathcal{Q})$ is a homomorphic image of a sublattice of $\mathrm{L_q}(\mathcal{K})$. This notion was introduced by Sapir [152], who showed that certain quasivarieties of semigroups are Q-universal. Moreover, if $\mathcal{K}$ is a Q-universal quasivariety, then

- $|L_q(\mathcal{K})| = 2^{\aleph_0}$,
- the free lattice $FL(\omega)$ is a sublattice of $L_q(\mathcal{K})$, whence in particular $L_q(\mathcal{K})$ satisfies no lattice identity.

For a good discussion of Q-universal quasivarieties, see Adams and Dziobiak [8].

The quasivarieties known to be Q-universal include deMorgan algebras, distributive p-algebras, Heyting algebras, the quasivariety generated by the modular lattice $\mathbf{M}_{3,3}$, MV-algebras, commutative rings with unity, all from [1], Kleene algebras [2], distributive lattices with a quantifier operation [3], pseudocomplemented semilattices [9], undirected graphs [6] (see also [110]), idempotent semigroups (bands) [7], Cantor algebras [160], and pointed abelian groups [137].

In this section we show that the quasivariety $\langle \mathbf{T}_3 \rangle$ is Q-universal (whence $\mathcal{M}$ is also). This strengthens Corollary 5.27, and requires a more detailed analysis of the quasivarieties $\langle \mathbf{P}_n \rangle$. We employ the following lemma of Adams and Dziobiak [1].

LEMMA 5.30. *Let N be a fixed countably infinite set and let $\mathcal{P}_{\mathrm{fin}}(N)$ denote the set of all finite subsets of it. Suppose $\mathcal{K}$ is a quasivariety of finite type that contains a family $(\mathbf{A}_X : X \in \mathcal{P}_{\mathrm{fin}}(N))$ of finite members satisfying the following conditions:*

- (P1) *$\mathbf{A}_\varnothing$ is a trivial member of $\mathcal{K}$;*
- (P2) *if $X = Y \cup Z$, then $\mathbf{A}_X \in \langle \mathbf{A}_Y \rangle \vee \langle \mathbf{A}_Z \rangle$;*
- (P3) *if $X \neq \varnothing$ and $\mathbf{A}_X \in \langle \mathbf{A}_Y \rangle$, then $X = Y$;*
- (P4) *if $\mathbf{A}_X \leq \mathbf{B} \times \mathbf{C}$ for finite $\mathbf{B}$ and $\mathbf{C} \in \mathcal{K}$, then there exist Y and Z such that $\mathbf{A}_Y \in \langle \mathbf{B} \rangle$, $\mathbf{A}_Z \in \langle \mathbf{C} \rangle$, and $X = Y \cup Z$.*

Then $\mathcal{K}$ is Q-universal. Moreover, the ideal lattice of $FL(\omega)$ is a sublattice of $L_q(\mathcal{K})$.

To apply the lemma to the quasivariety $\langle \mathbf{T}_3 \rangle$, take N to be the set of all prime numbers. For $X \in \mathcal{P}_{\mathrm{fin}}(N)$, let $\mathbf{A}_X = \mathbf{P}_n$ where $n = \prod\{q : q \in X\}$. Thus, as X varies, n runs over all positive square-free integers.

We also need to expand our list of quasi-equations. As in the proof of Lemma 5.28, the atoms of $\mathrm{Con}\,\mathbf{P}_n$ are

- α_{jk} given by $fx_j \equiv fx_k$ for $1 \leq j < k \leq n$,
- β_j given by $fx_j \equiv 0$ for all $1 \leq j \leq n$.

The congruences β_j, however, are not $\langle \mathbf{T}_3 \rangle$-congruences, since $\mathbf{T}_3$ satisfies $fx \approx 0 \to gx \approx x$, which is $\varepsilon_{\mathbf{T}_2, gx \equiv x}$. In other words, the atoms of $\mathrm{Con}_{\langle \mathbf{T}_3 \rangle}\,\mathbf{P}_n$ are just the congruences α_{jk} for $j < k$.

For $n \geq 2$ and $1 \leq j < k \leq n$, let ψ_{njk} denote the quasi-equation $\varepsilon_{\mathbf{P}_n, \alpha_{jk}}$, which is

$$fx_2 \approx gx_1 \;\&\; fx_3 \approx gx_2 \;\&\; \cdots \;\&\; fx_n \approx gx_{n-1} \;\&\; fx_1 \approx gx_n$$
$$\to fx_j \approx fx_k.$$

Note that ϕ_n is equivalent to ψ_{n12}, since $fx_2 \approx gx_1$ is part of the hypothesis.

LEMMA 5.31. *The quasi-equation ψ_{mst} is equivalent to ψ_{mdm} where* $d = \gcd(m, t - s)$.

PROOF. By circular symmetry we see that ψ_{mst} is equivalent to $\psi_{m(t-s)m}$. So consider ψ_{mem} with $1 \le e < m$. Whenever the hypotheses hold for a substitution $x_i \mapsto b_i$ in an algebra **B**, the conclusion yields $f(b_m) = f(b_{m+e}) = f(b_{m+2e}) = \cdots = f(b_{m+ke})$ for all $k \ge 0$, where the subscripts are taken modulo m. The least nonzero subscript we can obtain in this fashion, i.e., $f(b_{ke-\ell m})$ is of course $\gcd(e, m)$, with the conclusion $f(b_m) = f(b_d)$ where $d = \gcd(e, m)$. Thus ψ_{mem} implies ψ_{mdm}, while clearly ψ_{mdm} implies ψ_{mem} as in the first part of the argument.   □

Thus it suffices to consider the quasi-equations ψ_{mdm} with $1 \le d < m$ and $d \mid m$. Let $\widehat{\psi}_{md}$ denote ψ_{mdm} as a representative of its class.

A few elementary observations will prove useful.

LEMMA 5.32. *If $d \mid e \mid n$, then $\widehat{\psi}_{nd}$ implies $\widehat{\psi}_{ne}$.*

LEMMA 5.33. *$\widehat{\psi}_{nd}$ and $\widehat{\psi}_{ne}$ together imply $\widehat{\psi}_{ng}$ where $g = \gcd(d, e)$.*

PROOF. Assume that $\widehat{\psi}_{nd}$ and $\widehat{\psi}_{ne}$ both hold in an algebra $\mathbf{B} \in \mathcal{M}$, and that the hypotheses of the quasi-equations are satisfied for a substitution $x_i \mapsto b_i$. Then $f(b_n) = f(b_{kd+\ell e \bmod n})$ holds for all $k, \ell > 0$. The least such nonzero value of the subscript is the greatest common divisor of d and e.   □

LEMMA 5.34. *If $\mathcal{Q}$ satisfies $\widehat{\psi}_{nd}$ and $\mathcal{R}$ satisfies $\widehat{\psi}_{ne}$, then $\mathcal{Q} \vee \mathcal{R}$ satisfies $\widehat{\psi}_{n\ell}$ where $\ell = \mathrm{lcm}(d, e)$.*

PROOF. By Lemma 5.32, both $\mathcal{Q}$ and $\mathcal{R}$ satisfy $\widehat{\psi}_{n\ell}$.   □

Then we upgrade Theorem 5.26(4)–(5) to the version we need to apply Lemma 5.30.

LEMMA 5.35. *Let $m, n \ge 2$ and $d \mid m$, $e \mid n$.*

  (1) *The algebra $\mathbf{P}_n$ satisfies $\widehat{\psi}_{md}$ if and only if $n \nmid m$ or $n \mid d$.*
  (2) *$\widehat{\psi}_{md}$ implies $\widehat{\psi}_{ne}$ if and only if $n \mid m$ and $\gcd(n, d) \mid e$.*
  (3) *$\widehat{\psi}_{md}$ is equivalent to $\widehat{\psi}_{ne}$ if and only if $n = m$ and $d = e$.*

PROOF. For (1), consider substitutions $\sigma : \{x_1, \ldots, x_m\} \to \mathbf{P}_n$ such that the hypotheses of $\widehat{\psi}_{md}$ are satisfied. Let $a_1, \ldots, a_n$ denote the generators of $\mathbf{P}_n$. If $\sigma(x_1) \notin \{a_1, \ldots, a_n\}$, then as in the proof of Theorem 5.26(4) we obtain $\sigma(x_j) = 0$ for all j, whence the conclusion of $\widehat{\psi}_{md}$ holds as well. So we may assume that say $\sigma(x_1) = a_1$, by the circular symmetry of $\mathbf{P}_n$. Again arguing as in the proof of Theorem 5.26(4), we get $\sigma(x_{i_0+\ell}) = a_{1+\ell \bmod m}$, as in the proof of Theorem 5.26(4). Again using that the hypothesis of $\widehat{\psi}_{md}$ hold under the substitution, we must then have $\sigma(x_k) = a_{k \bmod n}$ for $1 \le k \le m$. In that event, the entire hypothesis holds under the substitution exactly when

$n \mid m$, and the conclusion $fx_n \approx fx_d$ holds exactly when $n \mid d$. So if $n \mid m$ and $n \nmid d$ then $\widehat{\psi}_{md}$ fails in $\mathbf{P}_m$ under this substitution, and otherwise it holds for all substitutions.

For (2), we first show that if the conditions fail then $\widehat{\psi}_{md}$ does not imply $\widehat{\psi}_{ne}$. If $n \nmid m$, then $\mathbf{P}_n$ satisfies $\widehat{\psi}_{md}$ but not $\widehat{\psi}_{ne}$. If $n \mid m$ but $\gcd(n,d) \nmid e$, then let $g = \gcd(n,d)$. In Con $\mathbf{P}_n$ form $\overline{\alpha}_g = \bigvee\{\alpha_{jk} : g \mid k-j\}$. Then $\mathbf{P}_n/\overline{\alpha}_g$ satisfies $\widehat{\psi}_{ng}$. Since $n \mid m$ and $g \mid d$, this implies it satisfies $\widehat{\psi}_{md}$. However, because $g \nmid e$, the algebra $\mathbf{P}_n/\overline{\alpha}_g$ fails $\widehat{\psi}_{ne}$. So again in this case, $\widehat{\psi}_{md}$ does not imply $\widehat{\psi}_{ne}$.

Conversely, assume that the conditions of part (2) hold, and that $\widehat{\psi}_{md}$ holds in an algebra $\mathbf{B} \in \mathcal{M}$. Suppose that the hypotheses of $\widehat{\psi}_{ne}$ are satisfied under a substitution $x_i \mapsto b_i$. As $n \mid m$, then the hypotheses of $\widehat{\psi}_{me}$ are also satisfied, whence the conclusion $f(b_m) = f(b_d)$ holds. But $b_m = b_n$, and as before $f(b_n) = f(b_{kg \bmod n})$ for all $k \geq 0$, where $g = \gcd(n,e)$. Since $g \mid e$, we conclude that $f(b_n) = f(b_e)$. Thus $\widehat{\psi}_{ne}$ holds in $\mathbf{B}$, as desired.

Part (3) follows immediately from (2). $\square$

LEMMA 5.36. *Let $e \mid n$ with $1 < e < n$. If the algebra $\mathbf{B} \in \langle \mathbf{T}_3 \rangle$ satisfies $\widehat{\psi}_{ne}$ but fails $\widehat{\psi}_{nk}$ for all $k < e$, then $\mathbf{P}_e \in \langle \mathbf{B} \rangle$.*

PROOF. For each divisor k of e with $1 \leq k < e$, the algebra $\mathbf{B}$ fails $\widehat{\psi}_{nk}$, so there exist elements $b_{k1}, \ldots, b_{kn} \in \mathbf{B}$ such that $f(b_{ki}) = g(b_{k,i-1})$ for all i (with subscripts taken mod n), but $f(b_{kn}) \neq f(b_{kd})$. However, $f(b_{ke}) = f(b_{kn})$. Also note that $f(b_{ki}) \neq 0$ for all i, because $\mathbf{T}_3$ satisfies $fx \approx 0 \to gx \approx x$.

Enumerate these proper divisors of e as $k_1, \ldots, k_t$. In the direct power $\mathbf{B}^{te}$ consider the elements

$$c_1 = (b_{k_1 1}, b_{k_1 2}, \ldots, b_{k_1 e}, \ldots, b_{k_t 1}, b_{k_t 2}, \ldots, b_{k_t e}),$$
$$c_2 = (b_{k_1 2}, \ldots, b_{k_1 e}, b_{k_1 1}, \ldots, b_{k_t 2}, \ldots, b_{k_t e}, b_{k_t 1}),$$
$$\cdots$$
$$c_e = (b_{k_1 e}, b_{k_1 1}, \ldots, b_{k_1, e-1}, \ldots, b_{k_t e}, b_{k_t 1}, \ldots, b_{k_t, e-1}).$$

Because $f(b_{kn}) = f(b_{ke})$ for all k, we have $f(c_i) = g(c_{i-1})$ for all i (with subscripts taken mod n).

Let us show that $f(c_1), \ldots, f(c_e)$ are all distinct. If perchance $f(c_i) = f(c_j)$ for some pair with $1 \leq i < j \leq e$, then by the construction $f(c_{s+i}) = f(c_{s+j})$ for all $s \geq 0$. Thus the subalgebra $\mathrm{Sg}(c_1, \ldots, c_e)$ satisfies $\widehat{\psi}_{e,t-s}$. As usual, this implies that $t - s$ divides e, and hence $t - s = k_u$ for some u. But this clearly fails, so we conclude that the elements $f(c_1), \ldots, f(c_e)$ are distinct.

This of course implies that $c_1, \ldots, c_e$ are distinct, and these elements generate a copy of $\mathbf{P}_e$. $\square$

THEOREM 5.37. *The quasivariety $\langle \mathbf{T}_3 \rangle$ is Q-universal.*

PROOF. We check that the algebras $\mathbf{A}_X$ defined above satisfy the conditions of Lemma 5.30.

Condition (P1) holds by definition.

For (P2), let X, Y, Z be finite subsets of n with $X = Y \cup Z$. Let $x = \prod X$, $y = \prod Y$, $z = \prod Z$ so that $\mathbf{A}_X = \mathbf{P}_x$, $\mathbf{A}_Y = \mathbf{P}_y$, $\mathbf{A}_Z = \mathbf{P}_z$. If $m \mid n$, then there is a natural congruence η_{nm} with $\mathbf{P}_n/\eta_{nm} \cong \mathbf{P}_m$. In our situation we have $y \mid x$, $z \mid x$ and $\eta_{xy} \wedge \eta_{xz} = \Delta$ in Con $\mathbf{P}_x$, since $X = Y \cup Z$ implies $\operatorname{lcm}(y, z) = x$. Thus $\mathbf{P}_x \leq \mathbf{P}_y \times \mathbf{P}_z$, whence $\mathbf{A}_X \in \langle \mathbf{A}_Y \rangle \vee \langle \mathbf{A}_Z \rangle$.

For (P3), assume $\varnothing \neq X \neq Y$, and again let $x = \prod X$ and $y = \prod Y$. If $y \nmid x$, then $\mathbf{P}_y$ satisfies ϕ_x while $\mathbf{P}_x$ does not satisfy ϕ_x. On the other hand, if $y \mid x$, then $\mathbf{P}_y$ satisfies $\psi_{x1(y+1)}$ but $\mathbf{P}_x$ does not satisfy $\psi_{x1(y+1)}$. Either way, $\mathbf{P}_x \notin \langle \mathbf{P}_y \rangle$, i.e., $\mathbf{A}_X \notin \langle \mathbf{A}_Y \rangle$.

For (P4), suppose $\mathbf{A}_X = \mathbf{P}_x \leq \mathbf{B} \times \mathbf{C}$. Then each quasi-equation $\widehat{\psi}_{xe}$ with $1 \leq e < x$ and $e \mid x$ fails in either $\mathbf{B}$ or $\mathbf{C}$. If all these quasi-equations fail in $\mathbf{B}$, say, then $\mathbf{P}_x \in \langle \mathbf{B} \rangle$ and we are done. Assuming that is not the case, choose d minimal such that $\widehat{\psi}_{xd}$ holds in $\mathbf{B}$. By Lemmas 5.32 and 5.33, it follows that $\widehat{\psi}_{xk}$ holds in $\mathbf{B}$ if and only if $d \mid k$.

Likewise, choose e minimal such that $\widehat{\psi}_{xe}$ holds in $\mathbf{C}$. Again, $\widehat{\psi}_{xk}$ holds in $\mathbf{C}$ if and only if $e \mid k$.

If say $d = 1$, then $\widehat{\psi}(n, d)$ holds in $\mathbf{B}$ for every $d \mid x$. Since every $\widehat{\psi}_{x,d}$ fails in the join, this implies that every $\widehat{\psi}_{x,d}$ fails in $\mathbf{C}$, and hence $\mathbf{P}_x \in \langle \mathbf{C} \rangle$. So we may assume $d > 1$, and likewise $e > 1$.

Similarly, we may assume that $\ell = \operatorname{lcm}(d, e) = x$, or else $\widehat{\psi}(x, \ell)$ holds in $\langle \mathbf{B} \rangle \vee \langle \mathbf{C} \rangle$, whence $\mathbf{P}_x \notin \langle \mathbf{B} \rangle \vee \langle \mathbf{C} \rangle$.

Now apply Lemma 5.36 to obtain $\mathbf{P}_d \in \langle \mathbf{B} \rangle$, and likewise $\mathbf{P}_e \in \langle \mathbf{C} \rangle$. Let Y be the set of prime factors of Y, and Z the prime factors of e. Since $\operatorname{lcm}(d, e) = x$ we have $X = Y \cup Z$, as desired. $\qquad \square$

COROLLARY 5.38. *The quasivariety* $\langle \mathbf{T}_3 \rangle$ *has* $2^{\aleph_0}$ *subquasivarieties, and the ideal lattice of* $\mathrm{FL}(\omega)$ *is a sublattice of* $\mathrm{L_q}(\langle \mathbf{T}_3 \rangle)$.

CHAPTER 6

1-Unary Algebras

6.1. 1-Unary Algebras With and Without 0

Now we turn our attention to an entirely different class of algebras. Our setting is either of two types of algebras.

- $\mathcal{N}$ is the variety of all algebras $\mathbf{T} = \langle T, f \rangle$ with one unary operation.
- $\mathcal{N}^0$ is the variety of all algebras $\mathbf{T} = \langle T, f, 0 \rangle$ with a unary operation and a constant satisfying $f0 \approx 0$.

We are particularly interested in the proper subvarieties of $\mathcal{N}$ and $\mathcal{N}^0$ that are determined by the equation $f^r x \approx f^s x$ for some pair $r < s$. Thus, for $0 \leq r < s$, we consider the following varieties.

- $\mathcal{N}_{r,s}$ is all algebras $\mathbf{T} = \langle T, f \rangle$ with one unary operation satisfying $f^r x \approx f^s x$.
- $\mathcal{N}^0_{r,s}$ is all algebras $\mathbf{T} = \langle T, f, 0 \rangle$ with a unary operation satisfying $f^r x \approx f^s x$ and a constant satisfying $f0 \approx 0$.

These are of course locally finite varieties. The variety $\mathcal{N}^0$ also has subvarieties $\langle f^r x \approx 0 \rangle$, but those are contained in $\mathcal{N}^0_{r,r+1}$.

The quasivarieties contained in $\mathcal{N}$ were described by Kartashov [96, 97, 98, 99], and the quasivarieties contained in $\mathcal{N}^0$ by Kartashov and Makaronov [102]. (See also Hyndman et al. [86].) Thus in this chapter we are primarily re-deriving previously known results. Our point in doing so is to illustrate in a simple setting how the general methodology reduces these investigations to rather straightforward calculations. In particular, for any fixed pair $r < s$, the lattices $\mathbf{L_q}(\mathcal{N}_{r,s})$ and $\mathbf{L_q}(\mathcal{N}^0_{r,s})$ are finite, and can be determined by hand whenever $s - r$ has few prime factors. We illustrate this in Sections 6.2 and 6.3.

In contrast, Viktor Gorbunov showed that the entire varieties $\mathcal{N}$ and $\mathcal{N}^0$ are Q-universal as quasivarieties [77]. This can be shown for $\mathcal{N}$ via Lemma 5.30. For a positive integer n, let $\mathbf{J}_{0,n}$ denote the cyclic 1-unary algebra of order n, so that $f^n(x) = x$ for all $x \in \mathbf{J}_{0,n}$. Let N be the set of all primes, and for a finite subset $X \subseteq N$ let $x = \prod X$, a square-free integer. It is straightforward to check that the algebras $\mathbf{A}_X = \mathbf{J}_{0,x}$ satisfy the properties

J. Hyndman, J. B. Nation, *The Lattice of Subquasivarieties of a Locally Finite Quasivariety*, CMS Books in Mathematics, https://doi.org/10.1007/978-3-319-78235-5_6

(P1)–(P4) of the lemma. To see that $\mathcal{N}^0$ is Q-universal, we use instead the algebras $\mathbf{A}_X^+ = \mathbf{J}_{0,x} + \mathbf{0}$ consisting of the disjoint union of a cycle and the fixed point 0.

Similar constructions to show the Q-universality of other types of algebras are given in Basheyeva *et al.* [25].

Now we turn to the subvarieties $\mathcal{N}_{r,s}^0$ and $\mathcal{N}_{r,s}$. Note that every finite algebra in $\mathcal{N}_{r,s}^0$ has a unique minimal generating set, *viz.*, $\{x \in \mathbf{T} : x \neq 0$ and $f(y) = x$ implies $y = x\}$. A similar statement applies to algebras in $\mathcal{N}_{r,s}$, omitting the reference to 0.

Let us develop some terminology to describe the finite algebras in these varieties. A *component* of $\mathbf{T}$ is a connected component of the graph of $\mathbf{T}$. The *core* of $\mathbf{T}$ is given by $\mathrm{core}(\mathbf{T}) = \bigcap_{k=1}^{\infty} \mathrm{range}(f^k) = \mathrm{range}(f^K)$ for sufficiently large K. Every component has a core which consists of a cycle or fixed point of f.

For $x \in T$, let $N(x) = \{y \in T : y \neq x$ and $f(y) = x\}$. (It is more convenient not to include x when it is a fixed point.) The *indegree* of x is given by $\mathrm{indeg}(x) = |N(x)|$. An element p is a *branch point* if $\mathrm{indeg}(p) > 1$.

The *branch* determined by an element $y \notin \mathrm{core}(\mathbf{T})$ is given by $B(y) = \{z \in T : f^k(z) = y$ for some $k \geq 0\}$; this includes y itself. The *length* of $B(y)$ is the number of elements in the longest chain $z, f(z), \ldots, f^{\ell-1}(z) = y$.

There are at least two kinds of folding retractions:

- If two branches meet at a branch point, you can fold the shorter one into the longer one.
- Any branch off the core can be folded into the core.

These can be formalized thusly. For the former, let $x \in T$ and $y_1, y_2 \in N(x) \setminus \mathrm{core}(\mathbf{T})$. Suppose $\ell_1 = \mathrm{length}(B(y_1)) \geq \mathrm{length}(B(y_2)) = \ell_2$. Choose a chain of maximal length $z_1, f(z_1), \ldots, f^{\ell_1-1}(z_1) = y_1$ in $B(y_1)$, so that $f^{\ell_1}(z_1) = x$. Then the following map is a retraction of $\mathbf{T}$ onto $\mathbf{T} \setminus B(y_2)$:

$$\varphi(t) = \begin{cases} t & \text{if } t \notin B(y_2), \\ f^{\ell_1-k}(z_1) & \text{if } t \in B(y_2) \text{ and } f^k(t) = x \text{ with } k \text{ minimal.} \end{cases}$$

For the latter type of folding, let $x \in \mathrm{core}(\mathbf{T})$ and $y \in N(x) \setminus \mathrm{core}(\mathbf{T})$, so that $f(y) = x \in \mathrm{core}(\mathbf{T})$. For each $k \geq 0$ there is a unique $z_k \in \mathrm{core}(\mathbf{T})$ such that $f^k(z_k) = x$. Thus we can define a retraction of $\mathbf{T}$ onto $\mathbf{T} \setminus B(y)$ by

$$\zeta(t) = \begin{cases} t & \text{if } t \notin B(y), \\ z_k & \text{if } t \in B(y) \text{ and } f^k(t) = x \text{ with } k \text{ minimal.} \end{cases}$$

These folding maps enable us to begin describing the quasicritical algebras in our varieties. An algebra $\mathbf{T}$ is not quasicritical exactly when there are proper subalgebras $\mathbf{S}_i \leq \mathbf{T}$ and endomorphisms $h_i : \mathbf{T} \to \mathbf{S}_i$ such that $\bigwedge_i \ker h_i = \Delta$. For 1-unary algebras, the folding retractions often serve as the desired maps h_i.

LEMMA 6.1. *Let* $\mathbf{T}$ *be a finite algebra in* $\mathbb{N}_{r,s}$ *or* $\mathbb{N}_{r,s}^0$.

(1) *If* $\mathbf{T}$ *has a point* x *with* $\operatorname{indeg}(x) > 2$, *then* $\mathbf{T}$ *is not quasicritical.*

(2) *If a component of* core$(\mathbf{T})$ *has more than one branch off it, then* $\mathbf{T}$ *is not quasicritical.*

(3) *If* core$(\mathbf{T})$ *has more than one component with a branch, then* $\mathbf{T}$ *is not quasicritical.*

(4) *If* $\mathbf{T}$ *has a branch* $B(x)$ *that contains a branch point* $y \neq x$, *then* $\mathbf{T}$ *is not quasicritical.*

PROOF. For (1), let $\mathbf{T}$ be an algebra containing an element x with $\operatorname{indeg}(x) > 2$. Let $N(x) = \{y_1, \ldots, y_m\}$ where $m \geq 3$, with y_1 chosen so that either $y_1 \in$ core$(\mathbf{T})$ or $B(y_1)$ has maximal length among the branches $B(y_j)$. Then $B(y_2)$ and $B(y_3)$ can be folded separately into the core or $B(y_1)$, as the case may be, and $\mathbf{T}$ is not quasicritical.

In parts (2) and (3), the offending branches can be folded separately into the core, showing that $\mathbf{T}$ is not quasicritical.

Part (4) requires a little more effort. Suppose that $\mathbf{T}$ contains a branch that consists of more than a single generator z and its images $f(z), \ldots, f^{\ell-1}(z)$. Then this branch is contained in a branch $B(x)$ with $x \in$ core$(\mathbf{T})$. Consider a branch point $y \in B(x)$ farthest out in the graph of $\mathbf{T}$, i.e., y is a branch point but $y \neq f^k(t)$ for any other branch point t and $k > 0$. By (1), we may assume that $\operatorname{indeg}(x) = 2$. Then there exist points z_1, z_2 and integers $k_1, k_2, c \geq 1$ such that

- $f^{k_j}(z_j) = y$ and $B(f^{k_j-1}(z_j))$ contains only $z_j, \ldots, f^{k_j-1}(z_j)$ for $j = 1, 2$,
- $f^c(y) = x \in$ core$(\mathbf{T})$ with c minimal.

The branch $B(f^{c-1}(y))$ may contain other elements not named. This configuration is illustrated in Figure 6.1. Without loss of generality $k_1 \geq k_2$, and so there is a retraction folding $B(f^{k_2-1}(z_2))$ into $B(f^{k_1-1}(z_1))$.

To show that $\mathbf{T}$ is not quasicritical, we need a second homomorphism ξ of $\mathbf{T}$ onto a subalgebra of $\mathbf{T}$ such that $\ker \xi$ is the identity on $B(f^{k_2-1}(z_2))$. Let C denote the component of $\mathbf{T}$ containing x.

(i) For $t \notin C$ define $\xi(t) = t$.

(ii) For each $t \in C$ we have $f^m(t) = x$ for some minimal m, and there is a unique $u_t \in$ core$(\mathbf{T})$ such that $f^m(u_t) = x$. If $t \in C \setminus B(f^{k_2-1}(z_2))$, define $\xi(t) = u_t$.

(iii) If $t \in B(f^{k_2-1}(z_2))$, then $t = f^j(z_2)$ for some j with $0 \leq j < k_2$. Define $\xi(t) = f^{c+j}(z_2)$.

On the component C, the map ξ shifts $B(f^{k_2-1}(z_2))$ forward by c and wraps all other elements around the core, so that the image $\xi(C)$ consists of its core and a single branch of length k_2. In case (iii) note that $f^{k_2-j}(t) = y$ and $f^c(y) = f^{c+k_2}(z_2) = x$ minimally. Thus $f^{k_2-j}(\xi(t)) = x = \xi(y)$. It is not hard to complete the check that ξ has the required properties. $\qquad \square$

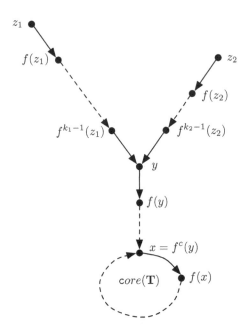

FIGURE 6.1. Configuration in the proof of Lemma 6.1(4)

Lemma 6.1 restricts the components of a quasicritical algebra. The candidates are these forms.

- **0** is a singleton $\{0\}$.
- **E** is a single fixed point $\{e\}$ with $f(e) = e$.
- $\mathbf{J}_{k,\ell}$ where $0 \le k < \ell$ is a cyclic algebra with a generator satisfying $f^k x = f^\ell x$.

Note $k = 0$ gives a cycle, while $k = k + 1$ gives a fixed point. Also $\mathbf{E} = \mathbf{J}_{0,1}$ but sometimes it is nice to distinguish this one. Let $\mathbf{S} + \mathbf{T}$ denote the disjoint union. An example is illustrated in Figure 6.2.

FIGURE 6.2. The unary algebra $\mathbf{J}_{1,2} + \mathbf{J}_{2,5}$

Lemma 6.1(3) says that a quasicritical algebra has at most one component with a branch, which has the following consequences.

COROLLARY 6.2. *If an algebra* $\mathbf{T}$ *in* $\mathcal{N}_{r,s}$ *or* $\mathcal{N}_{r,s}^0$ *has components* $\mathbf{J}_{k,\ell}$ *and* $\mathbf{J}_{k',\ell'}$ *with* $k > 0$ *and* $k' > 0$, *then* $\mathbf{T}$ *is not quasicritical.*

COROLLARY 6.3. *Every quasicritical algebra in $\mathcal{N}_{r,s}$ or $\mathcal{N}_{r,s}^0$ has the form* $\mathbf{J}_{k_1,\ell_1} + \cdots + \mathbf{J}_{k_m,\ell_m}$ *for some* $m \geq 1$, *where* $k_1 \leq r$ *and* $k_2 = \cdots = k_m = 0$ *and* $\ell_j - k_j | s - r$ *for each* j.

It remains to determine which of these sums is indeed quasicritical, remembering that quasicritical algebras (and subalgebras!) must contain 0 in $\mathcal{N}_{r,s}^0$. For a single component, this is easy.

LEMMA 6.4. (1) *Every algebra* $\mathbf{J}_{k,\ell}$ *in* $\mathcal{N}_{r,s}$ *with* $0 \leq k < \ell$ *and* $\ell - k | s - r$ *is quasicritical, except the 1-element algebra* $\mathbf{J}_{0,1} = \mathbf{E}$.
(2) *Every algebra* $\mathbf{J}_{k,k+1}$ *in* $\mathcal{N}_{r,s}^0$ *with* $0 < k$ *is quasicritical.*

PROOF. If $k = 0$ and $\ell > 1$, then $\mathbf{J}_{k,\ell}$ is a nontrivial algebra with no proper subalgebras, and hence quasicritical. If $k > 0$, then every proper subalgebra of $\mathbf{J}_{k,\ell}$ satisfies $f^{k-1}x \approx f^{\ell-1}x$, so again $\mathbf{J}_{k,\ell}$ is quasicritical. □

Turning to algebras with two or more components, it is convenient to first consider the case when $\mathbf{T}$ has a fixed point. In $\mathcal{N}_{r,s}^0$ we must exercise some care to distinguish between 0 and non-0 fixed points.

LEMMA 6.5. (1) *If an algebra* $\mathbf{T}$ *in* $\mathcal{N}_{r,s}$ *has at least two components, including a component* $\mathbf{J}_{k,k+1}$ *with* $k > 0$, *then* $\mathbf{T}$ *is not quasicritical.*
(2) *If an algebra* $\mathbf{T}$ *in* $\mathcal{N}_{r,s}^0$ *has at least two components, and the component containing 0 is* $\mathbf{J}_{k,k+1}$ *with* $k > 0$, *then* $\mathbf{T}$ *is not quasicritical.*

PROOF. In either case, let $\mathbf{T} = \mathbf{J}_{k,k+1} + \mathbf{R}$, where $\mathbf{R}$ denotes the remaining components. Then retractions onto $\mathbf{J}_{k,k+1}$ and $\mathbf{E} + \mathbf{R}$ give a subdirect decomposition of $\mathbf{T}$ into subalgebras. □

LEMMA 6.6. *If* $\mathbf{T}$ *in* $\mathcal{N}_{r,s}$ *or* $\mathcal{N}_{r,s}^0$ *contains a fixed point and has more than two components, then it is not quasicritical.*

PROOF. Assume $\mathbf{T} = \mathbf{U} + \mathbf{V} + \mathbf{W}$ where $\mathbf{U}$ contains a fixed point; if $\mathbf{T} \in \mathcal{N}_{r,s}^0$ we may suppose $0 \in \mathbf{U}$. Then retractions onto $\mathbf{U} + \mathbf{V}$ and $\mathbf{U} + \mathbf{W}$ give a subdirect decomposition of $\mathbf{T}$ into subalgebras. □

LEMMA 6.7. (1) *In* $\mathcal{N}_{r,s}$, *every algebra* $\mathbf{J}_{k,\ell} + \mathbf{E}$ *with* $\ell - k > 1$ *and* $\ell - k | s - r$ *is quasicritical. Also,* $\mathbf{E} + \mathbf{E}$ *is quasicritical.*
(2) *In* $\mathcal{N}_{r,s}^0$, *every algebra* $\mathbf{J}_{k,\ell} + \mathbf{0}$ *with* $\ell - k | s - r$ *is quasicritical. This includes* $\mathbf{E} + \mathbf{0}$.

These observations finish the cases with fixed points, including 0. So for the rest of this section we consider algebras in $\mathcal{N}_{r,s}$ that do not contain $\mathbf{J}_{k,k+1}$. Yet another easy lemma helps to reduce the number of cases.

LEMMA 6.8. *If* $\mathbf{T}$ *in* $\mathcal{N}_{r,s}$ *has components* $\mathbf{J}_{k,\ell}$ *and* $\mathbf{J}_{0,\ell'}$ *such that* $\ell' = \ell - k > 1$, *then* $\mathbf{T}$ *is not quasicritical.*

PROOF. Under the hypotheses, there are retractions mapping $\mathbf{J}_{0,\ell'}$ to $\mathbf{J}_{k,\ell}$ in separate ways, giving a subdirect decomposition of $\mathbf{T}$ into subalgebras.
□

Note that Lemma 6.8 includes the fact that $\mathbf{J}_{0,\ell} + \mathbf{J}_{0,\ell}$ is not quasicritical when $\ell > 1$.

When no component has a branch off its core, the generalization of Lemmas 6.6 and 6.8 is immediate.

LEMMA 6.9. (1) *The algebra* $\mathbf{J}_{0,\ell} + \mathbf{J}_{0,\ell'}$ *with* ℓ, $\ell' > 1$ *is quasicritical if and only if* $\ell \neq \ell'$.

(2) *More generally,* $\mathbf{T} = \mathbf{J}_{0,\ell_1} + \cdots + \mathbf{J}_{0,\ell_m}$ *with* $m > 1$ *and* $\ell_j > 1$ *for all* j *is quasicritical if and only if*

(i) $\ell_i \neq \ell_j$ *for* $i \neq j$,

(ii) *there do not exist* i, j, i', j' *with* $i \neq j$, $i' \neq j'$ *and* $j \neq j'$ *such that* $\ell_i \,|\, \ell_j$ *and* $\ell_{i'} \,|\, \ell_{j'}$,

(iii) *if* $\ell_{i_1}, \ldots, \ell_{i_n} \,|\, \ell_j$, *then* $\mathrm{lcm}(\ell_{i_1}, \ldots, \ell_{i_n}) < \ell_j$.

PROOF. If $\ell_i \,|\, \ell_j$ for some pair $i \neq j$, there is a retraction of $\mathbf{T}$ onto $\mathbf{T} \setminus \mathbf{J}_{0,\ell_j}$. The conditions ensure that the intersection of the kernels of two or more such retractions is not Δ. Part (2)(i) is just an important special case of (2)(ii). Note that the criterion of part (2)(ii) allows $i = i'$: if $\ell_i \,|\, \ell_j$ and $\ell_i \,|\, \ell_{j'}$ for $j \neq j'$, then $\mathbf{T}$ is not quasicritical. □

Thus, with respect to (ii), the algebra $\mathbf{J}_{0,2} + \mathbf{J}_{0,6} + \mathbf{J}_{0,10}$ is not quasicritical, as retractions onto $\mathbf{J}_{0,2} + \mathbf{J}_{0,6}$ and $\mathbf{J}_{0,2} + \mathbf{J}_{0,10}$ give a subdirect decomposition. Likewise, $\mathbf{J}_{0,2} + \mathbf{J}_{0,5} + \mathbf{J}_{0,6} + \mathbf{J}_{0,15}$ is not quasicritical, as is seen by using retractions onto $\mathbf{J}_{0,2} + \mathbf{J}_{0,5} + \mathbf{J}_{0,15}$ and $\mathbf{J}_{0,2} + \mathbf{J}_{0,5} + \mathbf{J}_{0,6}$.

With respect to (iii), $\mathbf{J}_{0,2} + \mathbf{J}_{0,3} + \mathbf{J}_{0,12}$ is quasicritical, but the algebra $\mathbf{J}_{0,4} + \mathbf{J}_{0,3} + \mathbf{J}_{0,12}$ is not. In both cases there are two retractions of $\mathbf{T}$ onto $\mathbf{T} \setminus \mathbf{J}_{0,12}$. In the latter case, the intersection of the kernels of those retractions is Δ, whereas in the former it is not.

Finally, we add a branch to the first component.

LEMMA 6.10. (1) *The algebra* $\mathbf{J}_{k,\ell} + \mathbf{J}_{0,\ell'}$ *in* $\mathcal{N}_{r,s}$ *with* $k > 0$, $\ell - k > 1$, *and* $\ell' > 1$ *is quasicritical if and only if* $\ell - k \nmid \ell'$.

(2) *More generally,* $\mathbf{J}_{k_1,\ell_1} + \mathbf{J}_{0,\ell_2} + \cdots + \mathbf{J}_{0,\ell_m}$ *in* $\mathcal{N}_{r,s}$ *with* $k_1 > 0$ *and* $m > 1$ *and* $\ell_1 - k_1 > 1$ *and* $\ell_j > 1$ *for* $j > 1$ *is quasicritical if and only if* $\ell_i \nmid \ell_j$ *for* $i, j > 1$ *and* $i \neq j$, *and* $\ell_1 - k_1 \nmid \ell_j$ *for all* $j > 1$.

PROOF. Now, in addition to a retraction onto $\mathbf{T} \setminus \mathbf{J}_{0,\ell_j}$ when $\ell_i \,|\, \ell_j$ or $\ell_1 - k_1 \,|\, \ell_j$, there is a retraction folding the branch of the first component into its core. □

Thus, by part (1), $\mathbf{J}_{3,5} + \mathbf{J}_{0,4}$ is not quasicritical, and indeed the retractions onto $\mathbf{J}_{3,5}$ and $\mathbf{J}_{0,2} + \mathbf{J}_{0,4}$ give a subdirect decomposition. However, in the terminology of part (2), it is allowed that $\ell_j \,|\, \ell_1 - k_1$ in a quasicritical algebra. For example, $\mathbf{J}_{3,7} + \mathbf{J}_{0,2}$ is quasicritical, since the only homomorphic images that are subalgebras $\mathbf{J}_{0,4} + \mathbf{J}_{0,2}$ and $\mathbf{J}_{0,2}$, with kernels that do not meet to Δ.

The preliminary calculations are done, and now we can combine them to classify the quasicritical algebras in $\mathcal{N}_{r,s}^0$ and $\mathcal{N}_{r,s}$. For the former, we need only Lemmas 6.1–6.7.

THEOREM 6.11. *The quasicritical algebras in* $\mathcal{N}_{r,s}^0$ *are*

(1) $\mathbf{J}_{k,k+1}$ *for* $0 < k \leq r$;
(2) $\mathbf{J}_{k,\ell} + \mathbf{0}$ *for* $0 \leq k \leq r$, $k < \ell$ *and* $\ell - k \mid s - r$.

THEOREM 6.12. *The quasicritical algebras in* $\mathcal{N}_{r,s}$ *are*

(1) $\mathbf{J}_{k,k+1}$ *for* $0 < k \leq r$;
(2) $\mathbf{J}_{k,\ell}$ *for* $0 \leq k \leq r$, $k + 1 < \ell$ *and* $\ell - k \mid s - r$;
(3) $\mathbf{E} + \mathbf{E}$;
(4) $\mathbf{J}_{0,\ell} + \mathbf{J}_{0,\ell'}$ *for* $0 < \ell$, $0 < \ell'$, $\ell \mid s - r$, $\ell' \mid s - r$ *and* $\ell \neq \ell'$;
(5) $\mathbf{J}_{0,\ell_1} + \cdots + \mathbf{J}_{0,\ell_m}$ *with* $m > 2$ *and* $0 < \ell_j$ *and* $\ell_j \mid s - r$ *for all* j, *subject to the conditions of* Lemma 6.9(2);
(6) $\mathbf{J}_{k,\ell} + \mathbf{J}_{0,\ell'}$ *for* $0 < k < \ell$, $0 < \ell'$, $\ell - k \mid s - r$, $\ell' \mid s - r$ *and* $\ell - k \nmid \ell'$;
(7) $\mathbf{J}_{k_1,\ell_1} + \mathbf{J}_{0,\ell_2} + \cdots + \mathbf{J}_{0,\ell_m}$ *with* $k_1 > 0$ *and* $m > 2$ *and* $\ell_1 - k_1 > 1$ *and* $\ell_j > 1$ *for* $j > 1$ *and* $\ell_1 - k_1 \mid s - r$ *and* $\ell_j \mid s - r$ *for* $j > 1$, *subject to the conditions of* Lemma 6.10(2).

While the conditions for two or more components look ominous, they are not that bad. If a quasicritical algebra has more than one component, then at most one can have a branch. Then different conditions apply depending on whether one or none has a branch. Note that the algebras $\mathbf{J}_{k,\ell} + \mathbf{E}$ of Lemma 6.7(1) are included in parts (4) and (6) of Theorem 6.12.

COROLLARY 6.13. *If* $\mathbf{T} \in \mathcal{N}_{r,s}^0$ *is quasicritical, then* $\mathbf{T}$ *is 1-generated.*

COROLLARY 6.14. *If* $\mathbf{T} \in \mathcal{N}_{r,s}$ *is quasicritical, then* $\mathbf{T}$ *is generated by at most* $2^m + 1$ *elements, where* m *is the number of factors in the prime factorization of* $s - r$.

It follows that each $\mathcal{N}_{r,s}^0$ or $\mathcal{N}_{r,s}$ contains only finitely many subquasivarieties. In general, whenever there is a bound N on the number of generators of quasicritical algebras in a quasivariety $\mathcal{Q}$, then the subquasivarieties of $\mathcal{Q}$ are determined by quasi-equations in at most N variables, and indeed $\mathsf{L}_q(\mathcal{Q})$ is determined by the free algebra $\mathcal{F}_{\mathcal{Q}}(N)$, as in say [19]. In our case, the quasicritical algebras in $\mathcal{N}_{r,s}^0$ are 1-generated, while for $\mathcal{N}_{r,s}$ this bound is the maximum number of components in a quasicritical algebra, which is a function of $s - r$. Writing $\mu(s - r)$ to denote it, the bound $\mu(s - r) \leq 2^m + 1$ of Corollary 6.14 is pretty bad. A few small values are easy to find. We have $\mu(s - r) = 2$ if the prime factorization is $s - r = p^k$ for any $k \geq 0$, or if $s - r = pq$. Likewise, $\mu(s - r) = 3$ if $s - r = p^k q^\ell$ with $k > 1$ or if $s - r = p_1 p_2 p_3$. On the other hand, it follows from Lemma 6.12(5) that for distinct primes a lower bound is given by $\mu(p_1 \cdots p_n) \geq \binom{n}{\lfloor n/2 \rfloor}$.

Now let us consider the embedding relations between quasicritical algebras, and their subdirect decompositions into quasicritical factors. Remember that $\mathbf{S} \leq \mathbf{T}$ implies $\langle \mathbf{S} \rangle \leq \langle \mathbf{T} \rangle$, while $\mathbf{S} \leq \mathbf{T}_1 \times \cdots \times \mathbf{T}_m$ implies $\langle \mathbf{S} \rangle \leq \langle \mathbf{T} \rangle \vee \cdots \vee \langle \mathbf{T}_m \rangle$; the latter includes $\langle \mathbf{S} \rangle \leq \langle \mathbf{T} \rangle$ whenever $\mathbf{S} \leq \mathbf{T}^m$. Moreover, every finite algebra is a subdirect product of its quasicritical quotients. For any locally finite quasivariety $\mathcal{Q}$ of finite type, the quasivariety lattice $L_q(\mathcal{Q})$ is join-generated by the quasivarieties $\langle \mathbf{T} \rangle$ with $\mathbf{T} \in \mathcal{Q}$ quasi-critical, subject to these inclusions and join dependencies.

Analysis of the subdirect decomposition of quasicritical algebras in $\mathcal{N}_{r,s}^0$ or $\mathcal{N}_{r,s}$ is based on the following elementary observations.

LEMMA 6.15. *The algebra $\mathbf{J}_{k,\ell}$ has homomorphic images isomorphic to*

(1) $\mathbf{J}_{k',\ell-k+k'}$ *for* $0 \leq k' \leq k$,
(2) $\mathbf{J}_{k,k+(\ell-k)/d}$ *for* $d \mid \ell - k$,
(3) $\mathbf{J}_{k',k'+(\ell-k)/d}$ *for* $0 \leq k' \leq k$ *and* $d \mid \ell - k$.

PROOF. The first congruence folds part or all of the branch into the core. The second wraps the core into a smaller one. The third is the join of the first two. □

First consider embeddings and subdirect decompositions in $\mathcal{N}_{r,s}^0$.

LEMMA 6.16. *The quasicritical algebras in $\mathcal{N}_{r,s}^0$ satisfy the following embedding relations.*

(1) $\mathbf{J}_{k,k+1} \leq \mathbf{J}_{k',k'+1}$ *iff* $k \leq k'$.
(2) $\mathbf{J}_{k,\ell} + \mathbf{0} \leq \mathbf{J}_{k',\ell'} + \mathbf{0}$ *iff* $k \leq k'$ *and* $\ell - k = \ell' - k'$.

Note that, because of the constant, $\mathbf{J}_{k,k+1} \not\leq \mathbf{J}_{k,k+1} + \mathbf{0}$. Indeed, the quasivarieties $\langle \mathbf{J}_{k,k+1} \rangle$ and $\langle \mathbf{J}_{k,k+1} + \mathbf{0} \rangle$ are incomparable in $L_q(\mathcal{N}_{r,s}^0)$: the former satisfies $f^k x \approx 0$, while the latter satisfies $fx \approx 0 \rightarrow x \approx 0$.

LEMMA 6.17. *The quasicritical algebras in $\mathcal{N}_{r,s}^0$ have these subdirect decompositions.*

(1) $\mathbf{J}_{k,k+1}$ *is subdirectly irreducible for* $k > 0$.
(2) $\mathbf{J}_{0,\ell} + \mathbf{0}$ *is subdirectly irreducible if* ℓ *is 1 or a prime power.*
(3) *Else, whenever* $\ell - k = \mathrm{lcm}(q_1, \ldots, q_m)$, *then*

$$(i) \qquad \mathbf{J}_{0,\ell} + \mathbf{0} \leq \prod_{i=1}^{m}(\mathbf{J}_{0,q_i} + \mathbf{0}) \text{ for } k = 0,$$

$$(ii) \qquad \mathbf{J}_{k,\ell} + \mathbf{0} \leq \mathbf{J}_{k,k+1} \times \prod_{i=1}^{m}(\mathbf{J}_{0,q_i} + \mathbf{0}) \text{ for } k > 0,$$

$$(iii) \qquad \mathbf{J}_{k,\ell} + \mathbf{0} \leq (\mathbf{J}_{k,k+q_1} + \mathbf{0}) \times \prod_{i=2}^{m}(\mathbf{J}_{0,q_i} + \mathbf{0}) \text{ for } k > 0.$$

These decompositions are irredundant so long as $q_i \nmid q_j$ for $i \neq j$; decompositions of type (iii) *are also irredundant when $q_1 \mid q_n$ and $q_1 < q_n$ for one or more $n > 1$, and $q_i \nmid q_j$ for all pairs $i \neq j$ with $i > 1$.*

A few examples are in order. For the case with $k = 0$, we have $\mathbf{J}_{0,20} + \mathbf{0} \leq (\mathbf{J}_{0,q_1} + \mathbf{0}) \times (\mathbf{J}_{0,q_2} + \mathbf{0})$ irredundantly with quasicritical factors for the sets $\{q_1, q_2\} = \{4, 5\}$ or $\{4, 10\}$. Note that while the congruences giving the factors $\mathbf{J}_{0,5} + \mathbf{0}$ and $\mathbf{J}_{0,10} + \mathbf{0}$ are comparable, the two algebras generate incomparable quasivarieties: the former, but not the latter, satisfies $f^5 x \approx x$, while the latter, but not the former, satisfies $f^5 x \approx x \to x \approx 0$.

Similarly, $\mathbf{J}_{0,30} + \mathbf{0} \leq \prod_i (\mathbf{J}_{0,q_i} + \mathbf{0})$ irredundantly with quasicritical factors for the sets $\{2,3,5\}$, $\{2,15\}$, $\{3,10\}$, $\{5,6\}$, $\{6,10\}$, $\{6,15\}$, and $\{10,15\}$.

For an example with $k > 0$, we have

$$\mathbf{J}_{3,7} + \mathbf{0} \leq \mathbf{J}_{3,4} \times (\mathbf{J}_{0,4} + \mathbf{0}),$$
$$\mathbf{J}_{3,7} + \mathbf{0} \leq (\mathbf{J}_{3,4} + \mathbf{0}) \times (\mathbf{J}_{0,4} + \mathbf{0}),$$
$$\mathbf{J}_{3,7} + \mathbf{0} \leq (\mathbf{J}_{3,5} + \mathbf{0}) \times (\mathbf{J}_{0,4} + \mathbf{0}).$$

Note that the second and third decompositions fit the pattern of part (3)(iii) with $q_1 \mid q_2$ properly. The quasivarieties $\langle \mathbf{J}_{3,4} \rangle$ and $\langle \mathbf{J}_{3,4} + \mathbf{0} \rangle$ are incomparable, and the quasivariety inclusions obtained from all three decompositions should be included in the description of $\mathrm{L}_q(\mathcal{N}_{3,7}^0)$:

$$\langle \mathbf{J}_{3,7} + \mathbf{0} \rangle \leq \langle \mathbf{J}_{3,4} \rangle \vee \langle \mathbf{J}_{0,4} + \mathbf{0} \rangle,$$
$$\langle \mathbf{J}_{3,7} + \mathbf{0} \rangle \leq \langle \mathbf{J}_{3,4} + \mathbf{0} \rangle \vee \langle \mathbf{J}_{0,4} + \mathbf{0} \rangle,$$
$$\langle \mathbf{J}_{3,7} + \mathbf{0} \rangle \leq \langle \mathbf{J}_{3,4} + \mathbf{0} \rangle \vee \langle \mathbf{J}_{0,4} + \mathbf{0} \rangle.$$

The preceding examples concern a particular quasicritical algebra. Section 6.2 applies Lemmas 6.16 and 6.17 to all the quasicritical algebras in $\mathcal{N}_{r,s}^0$ for some small values of r and s, yielding enough information to completely determine $\mathrm{L}_q(\mathcal{N}_{r,s}^0)$ for those quasivarieties.

Now we turn to embeddings and subdirect decompositions in $\mathcal{N}_{r,s}$.

LEMMA 6.18. *The quasicritical algebras in $\mathcal{N}_{r,s}$ satisfy the following embedding relations.*

(1) $\mathbf{J}_{k,\ell} \leq \mathbf{J}_{k',\ell'}$ *iff* $k \leq k'$ *and* $\ell - k = \ell' - k'$.

(2) *More generally,* $\sum_I \mathbf{J}_{k_i,\ell_i} \leq \sum_J \mathbf{J}_{m_j,n_j}$ *if and only if there is a one-to-one map* $\pi : I \to J$ *such that* $\mathbf{J}_{k_i,\ell_i} \leq \mathbf{J}_{m_{\pi(i)},n_{\pi(i)}}$ *holds for all i, that is,* $k_i \leq m_{\pi(i)}$ *and* $\ell_i - k_i = n_{\pi(i)} - m_{\pi(i)}$.

The next lemma is a direct analogue of Lemma 6.17 for quasicritical algebras in $\mathcal{N}_{r,s}$ that have one component, or one component plus $\mathbf{E}$.

LEMMA 6.19. *The following statements about subdirect decompositions hold for quasicritical algebras in $\mathcal{N}_{r,s}$.*

(1) $\mathbf{J}_{k,k+1}$ *is subdirectly irreducible for $k > 0$.*

(2) $\mathbf{J}_{0,\ell}$ *is subdirectly irreducible if $\ell > 1$ is a prime power.*

(3) *Else, whenever* $\ell - k = \mathrm{lcm}(q_1, \dots, q_m)$, *then*

(i) $\displaystyle \mathbf{J}_{0,\ell} \leq \prod_{i=1}^{m} \mathbf{J}_{0,q_i}$ *for $k = 0$,*

(ii) $\displaystyle \mathbf{J}_{k,\ell} \leq \mathbf{J}_{k,k+1} \times \prod_{i=1}^{m} \mathbf{J}_{0,q_i}$ *for $k > 0$,*

(iii) $\displaystyle \mathbf{J}_{k,\ell} \leq \mathbf{J}_{k,k+q_1} \times \prod_{i=2}^{m} \mathbf{J}_{0,q_i}$ *for $k > 0$.*

(4) $\mathbf{J}_{0,\ell} + \mathbf{E}$ *is subdirectly irreducible if ℓ is* 1 *or a prime power.*

(5) *Else, whenever $\ell - k = \mathrm{lcm}(q_1, \ldots, q_m)$, then*

$$\text{(i)} \qquad \mathbf{J}_{0,\ell} + \mathbf{E} \le \prod_{i=1}^{m} (\mathbf{J}_{0,q_i} + \mathbf{E}) \ \text{for } k = 0,$$

$$\text{(ii)} \qquad \mathbf{J}_{k,\ell} + \mathbf{E} \le \mathbf{J}_{k,k+1} \times \prod_{i=1}^{m} (\mathbf{J}_{0,q_i} + \mathbf{E}) \ \text{for } k > 0,$$

$$\text{(iii)} \qquad \mathbf{J}_{k,\ell} + \mathbf{E} \le (\mathbf{J}_{k,k+q_1} + \mathbf{E}) \times \prod_{i=2}^{m} (\mathbf{J}_{0,q_i} + \mathbf{E})$$

$$\text{for } k > 0.$$

The decompositions in parts (3) *and* (5) *are irredundant so long as $q_i \nmid q_j$ for $i \ne j$; decompositions of type* (iii) *are also irredundant when $q_1 \mid q_n$ and $q_1 < q_n$ for one or more $n > 1$, and $q_i \nmid q_j$ for all pairs $i \ne j$ with $i > 1$.*

For quasicritical algebras with two or more nontrivial components, the statements are less straightforward. Nonetheless, there is an algorithm to find the irredundant subdirect decompositions of a given quasicritical algebra in $\mathcal{N}_{r,s}$.

Fix a quasicritical algebra $\mathbf{T}$ in $\mathcal{N}_{r,s}$. Our task is to classify the sets of images $\mathbf{T}/\varphi_j$ that arise when $\Delta = \bigwedge_j \varphi_j$ irredundantly in $\mathrm{Con}\,\mathbf{T} = \mathrm{Con}_{\mathcal{N}_{r,s}}\mathbf{T}$. Parts (1)–(3) of Lemma 6.19 cover the case when $\mathbf{T} \cong \mathbf{J}_{k,\ell}$ has a single component. On the other hand, when an algebra has several components, say $\mathbf{T} = \mathbf{C}_1 + \cdots + \mathbf{C}_m$, a congruence $\varphi = \bigwedge_k \varphi_k$ is Δ if and only if

(i) the restriction of φ to each $\mathbf{C}_i$ is Δ,

(ii) φ contains no pair (x, y) with x, y in different components.

The following steps outline a way to systematically find all such combinations for a quasicritical algebra $\mathbf{T} = \mathbf{C}_1 + \cdots + \mathbf{C}_m$.

(1) List the homomorphic images of each component $\mathbf{C}_i$.

(2) Combine these to find the quasicritical proper homomorphic images of $\mathbf{T}$, which are of the form $\mathbf{D}_k = \mathbf{D}_{k1} + \cdots + \mathbf{D}_{kn_k}$ where every $\mathbf{D}_{kj}$ is an image of some $\mathbf{C}_i$, and for each i there exists j such that $\mathbf{D}_{kj}$ is an image of $\mathbf{C}_i$. Keep track of the kernels φ_k of the corresponding homomorphisms.

(3) Find the subsets S of these kernels such that $\bigwedge_{k \in S} \varphi_k = \Delta$ irredundantly in $\mathrm{Con}\,\mathbf{T}$, so that (i) and (ii) above are satisfied.

The results of applying the algorithm to quasicritical algebras with two components, where the number of combinations is manageable, can be summarized as follows.

LEMMA 6.20. *The quasicritical algebras $\mathbf{J}_{k,\ell} + \mathbf{J}_{0,b}$ in $\mathcal{N}_{r,s}$ have these subdirect decompositions.*

(1) $\mathbf{J}_{0,a} + \mathbf{J}_{0,b} \le_s \prod_{i=1}^{m}(\mathbf{J}_{0,q_i} + \mathbf{J}_{0,r_i})$ *if and only if $\mathrm{lcm}(q_1, \ldots, q_m) = a$ and $\mathrm{lcm}(r_1, \ldots, r_m) = b$.*

(2) $\mathbf{J}_{0,a} + \mathbf{J}_{0,b} \leq_s \prod_{i=1}^m \mathbf{J}_{0,d_i} \times \prod_{j=1}^n (\mathbf{J}_{0,q_j} + \mathbf{J}_{0,r_j})$ *if and only if* $\mathrm{lcm}(d_1,\ldots,d_m,q_1,\ldots,q_n) = a$ *and* $\mathrm{lcm}(d_1,\ldots,d_m,r_1,\ldots,r_n) = b$.

(3) $\mathbf{J}_{k,k+a} + \mathbf{J}_{0,b} \leq_s \prod_{i=1}^m \mathbf{J}_{c_i,c_i+d_i} \times \prod_{j=1}^n (\mathbf{J}_{e_j,e_j+q_j} + \mathbf{J}_{0,r_j})$ *for* $k \geq 0$ *if and only if* $\max(c_1,\ldots,c_m,e_1,\ldots,e_n) = k$ *and* $\mathrm{lcm}(d_1,\ldots,d_m,q_1,\ldots,q_n) = a$ *and* $\mathrm{lcm}(d_1,\ldots,d_m,r_1,\ldots,r_n) = b$.

We illustrate the procedure with examples of increasing complexity.

The algebra $\mathbf{S} = \mathbf{J}_{0,2} + \mathbf{J}_{0,5}$ has components that are prime cycles. The only images of $\mathbf{J}_{0,p}$ for p prime are itself and $\mathbf{E}$. Thus $\mathbf{S}$ has five congruences, with the images $\mathbf{J}_{0,2} + \mathbf{J}_{0,5}$, $\mathbf{E} + \mathbf{J}_{0,5}$, $\mathbf{J}_{0,2} + \mathbf{E}$, $\mathbf{E} + \mathbf{E}$ and $\mathbf{E}$. The only combination that gives a subdirect decomposition is $\mathbf{J}_{0,2} + \mathbf{J}_{0,5} \leq (\mathbf{E} + \mathbf{J}_{0,5}) \times (\mathbf{J}_{0,2} + \mathbf{E})$.

More generally, whenever p_i for $1 \leq i \leq m$ are distinct primes, then $\sum_{i=1}^m \mathbf{J}_{0,p_i} \leq \prod_{i=1}^m (\mathbf{J}_{0,p_i} + \mathbf{E})$ is the only subdirect decomposition of $\sum_i \mathbf{J}_{0,p_i}$ into quasicritical factors.

The algebra $\mathbf{T} = \mathbf{J}_{0,4} + \mathbf{J}_{0,5}$ is only slightly different. When q is a prime power, then $\mathbf{J}_{0,q}$ is subdirectly irreducible. For example, the first component $\mathbf{J}_{0,4}$ of $\mathbf{T}$ has images $\mathbf{J}_{0,4}$, $\mathbf{J}_{0,2}$, and $\mathbf{E}$. Since $\mathbf{J}_{0,4}$ is subdirectly irreducible, it must be used as a component in some factor of any decomposition. Thus we obtain $\mathbf{J}_{0,4} + \mathbf{J}_{0,5} \leq (\mathbf{J}_{0,4} + \mathbf{E}) \times (\mathbf{E} + \mathbf{J}_{0,5})$ and $\mathbf{J}_{0,4} + \mathbf{J}_{0,5} \leq (\mathbf{J}_{0,4} + \mathbf{E}) \times (\mathbf{J}_{0,2} + \mathbf{J}_{0,5})$.

Next consider $\mathbf{U} = \mathbf{J}_{0,2} + \mathbf{J}_{0,15}$. Its homomorphic images are algebras $\mathbf{J}_{0,a} + \mathbf{J}_{0,b}$ with $a \in \{1,2\}$ and $b \in \{1,3,5,15\}$, and $\mathbf{E}$. Moreover, $\mathbf{J}_{0,2} + \mathbf{J}_{0,15} \leq (\mathbf{J}_{0,a} + \mathbf{J}_{0,b}) \times (\mathbf{J}_{0,c} + \mathbf{J}_{0,d})$ holds exactly when $\mathrm{lcm}(a,c) = 2$ and $\mathrm{lcm}(b,d) = 15$. Up to symmetry, we obtain proper subdirect decompositions of $\mathbf{U}$ for $(a,b,c,d) = (2,1,1,15)$, $(2,3,1,5)$, $(2,3,1,15)$, $(2,5,1,3)$, $(2,5,1,15)$, and $(2,3,2,5)$. There are also irredundant subdirect decompositions with three factors:

$$\mathbf{U} \leq (\mathbf{J}_{0,a} + \mathbf{J}_{0,b}) \times (\mathbf{J}_{0,c} + \mathbf{J}_{0,d}) \times (\mathbf{J}_{0,e} + \mathbf{J}_{0,g})$$

for the 6-tuples $(a,b,c,d,e,g) = (2,1,1,3,1,5)$, $(2,1,1,3,2,5)$, and $(2,1,2,3,1,5)$. The decompositions with three factors are irrelevant for our purpose of determining the structure of $L_q(\mathcal{N}_{0,30})$, since they give join covers that do not belong to the E-basis (see Corollary 4.11). For example, with the first 6-tuple,

$$\langle \mathbf{J}_{0,2} + \mathbf{J}_{0,15} \rangle \leq \langle \mathbf{J}_{0,2} + \mathbf{J}_{0,3} \rangle \vee \langle \mathbf{J}_{0,2} + \mathbf{J}_{0,5} \rangle$$
$$< \langle \mathbf{J}_{0,2} + \mathbf{E} \rangle \vee \langle \mathbf{E} + \mathbf{J}_{0,3} \rangle \vee \langle \mathbf{E} + \mathbf{J}_{0,5} \rangle.$$

(The last inclusion is strict because $\langle \mathbf{J}_{0,2} + \mathbf{J}_{0,3} \rangle \vee \langle \mathbf{J}_{0,2} + \mathbf{J}_{0,5} \rangle$ satisfies the quasi-equation $fx \approx x \to x \approx y$.) Hence these triple decompositions, though valid, may be omitted from our basis for $L_q(\mathcal{N}_{0,30})$, as they can be derived from those with two factors.

Let $\mathbf{V} = \mathbf{J}_{0,2} + \mathbf{J}_{0,10}$. Then $\mathbf{J}_{0,2}$ is a homomorphic image of $\mathbf{V}$. The subdirect representations involving the image $\mathbf{J}_{0,2}$ are

$$\mathbf{J}_{0,2} + \mathbf{J}_{0,10} \leq \mathbf{J}_{0,2} \times (\mathbf{E} + \mathbf{J}_{0,5}),$$

$$\mathbf{J}_{0,2} + \mathbf{J}_{0,10} \leq \mathbf{J}_{0,2} \times (\mathbf{E} + \mathbf{J}_{0,10}),$$
$$\mathbf{J}_{0,2} + \mathbf{J}_{0,10} \leq \mathbf{J}_{0,2} \times (\mathbf{J}_{0,2} + \mathbf{J}_{0,5}).$$

The first two give proper join covers in $L_q(\mathcal{N}_{0,10})$, but since $\mathbf{J}_{0,2} \leq \mathbf{J}_{0,2} + \mathbf{J}_{0,5}$ the last one translates into the quasivariety inclusion

$$\langle \mathbf{J}_{0,2} + \mathbf{J}_{0,10} \rangle \leq \langle \mathbf{J}_{0,2} \rangle \vee \langle \mathbf{J}_{0,2} + \mathbf{J}_{0,5} \rangle = \langle \mathbf{J}_{0,2} + \mathbf{J}_{0,5} \rangle.$$

It is not hard to see that the remaining subdirect representations of $\mathbf{V}$ give inclusions that are refined by these, and hence not part of the E-basis.

The same argument shows that if a and b are relatively prime, then $\langle \mathbf{J}_{0,a} + \mathbf{J}_{0,ab} \rangle \leq \langle \mathbf{J}_{0,a} + \mathbf{J}_{0,b} \rangle$.

Now consider $\mathbf{W} = \mathbf{J}_{2,4} + \mathbf{J}_{0,5}$, which has a component $\mathbf{J}_{k,\ell}$ with $k > 0$. The images of the first component are $\mathbf{E}$, $\mathbf{J}_{1,2}$, $\mathbf{J}_{2,3}$, $\mathbf{J}_{0,2}$, $\mathbf{J}_{1,3}$, and $\mathbf{J}_{2,4}$. However, we must use either $\mathbf{J}_{2,3}$ or $\mathbf{J}_{2,4}$ as a component to get the branch. Thus we want the decompositions

$$\mathbf{J}_{2,4} + \mathbf{J}_{0,5} \leq \mathbf{J}_{2,3} \times (\mathbf{J}_{0,2} + \mathbf{J}_{0,5}),$$
$$\mathbf{J}_{2,4} + \mathbf{J}_{0,5} \leq (\mathbf{J}_{2,4} + \mathbf{E}) \times (\mathbf{E} + \mathbf{J}_{0,5}).$$

Replacing $\mathbf{J}_{0,2} + \mathbf{J}_{0,5}$ in the first decomposition by $(\mathbf{J}_{0,2} + \mathbf{E}) \times (\mathbf{E} + \mathbf{J}_{0,5})$ gives another subdirect representation, but again one not needed for the E-basis since $\langle \mathbf{J}_{0,2} + \mathbf{J}_{0,5} \rangle < \langle \mathbf{J}_{0,2} + \mathbf{E} \rangle \vee \langle \mathbf{E} + \mathbf{J}_{0,5} \rangle$ in the lattice of quasivarieties.

It is worthwhile to record some of the patterns we have seen in these examples.

LEMMA 6.21. *The following subdirect decompositions hold for quasicritical algebras in* $\mathcal{N}_{r,s}$.

(1) $\mathbf{J}_{0,\ell} + \mathbf{J}_{0,n} \leq (\mathbf{J}_{0,\ell} + \mathbf{E}) \times \mathbf{J}_{0,n}$ *if* $n \mid \ell$.
(2) $\mathbf{J}_{k,\ell} + \mathbf{J}_{0,n} \leq (\mathbf{J}_{k,\ell} + \mathbf{E}) \times \mathbf{J}_{0,n}$ *if* $n \mid (\ell - k)$ *for* $k \geq 0$.
(3) $\mathbf{J}_{k,\ell} + \mathbf{J}_{0,n} \leq (\mathbf{J}_{k,\ell} + \mathbf{E}) \times (\mathbf{E} + \mathbf{J}_{0,n})$ *for* $k \geq 0$.
(4) $\mathbf{J}_{k,\ell} + \mathbf{J}_{0,n} \leq \mathbf{J}_{k,k+1} \times (\mathbf{J}_{0,\ell-k} + \mathbf{J}_{0,n})$ *for* $k \geq 0$.

Section 6.3 shows how these methods may be applied to obtain enough information to determine $L_q(\mathcal{N}_{r,s})$ for small values of $s - r$.

Let us also note some information relevant to the structure of $L_q(\mathcal{N}_{r,s})$ that is implicit in the above discussion

If $\mathbf{T} = \mathbf{J}_{k,\ell} + \mathbf{J}_{0,n}$ is quasicritical, then the quasivariety $\langle \mathbf{T} \rangle$ is join irreducible, so $\langle \mathbf{J}_{k,\ell} \rangle \vee \langle \mathbf{J}_{0,n} \rangle < \langle \mathbf{T} \rangle$. If $\ell - k > 1$ and $n > 1$, then in fact the join $\langle \mathbf{J}_{k,\ell} \rangle \vee \langle \mathbf{J}_{0,n} \rangle$ is not generated by a single algebra.

We have observed that $\mathbf{J}_{0,a} + \mathbf{J}_{0,b} \leq (\mathbf{E} + \mathbf{J}_{0,a}) \times (\mathbf{E} + \mathbf{J}_{0,b})$. Indeed, it is not hard to see that $\langle \mathbf{E} + \mathbf{J}_{0,a} \rangle \vee \langle \mathbf{E} + \mathbf{J}_{0,b} \rangle = \langle \mathbf{E} + \mathbf{J}_{0,a} + \mathbf{J}_{0,b} \rangle$.

6.2. Some Quasivarieties $\mathcal{N}_{r,s}^0$

In this section we illustrate Theorem 6.11 and Lemmas 6.16 and 6.17 for quasivarieties $\mathcal{N}_{r,s}^0$. The following section similarly illustrates Theorem 6.12 and Lemmas 6.18, 6.19, and 6.20 for quasivarieties $\mathcal{N}_{r,s}$. In each case, this is the information needed to compute the lattice of quasivarieties $L_q(\mathcal{N}_{r,s}^0)$

or $L_q(\mathcal{N}_{r,s})$. These lattices tend to get large, so we only draw a few of the smaller ones. In view of Corollary 4.11, only subdirect decompositions that give join covers in the E-basis of $L_q(\mathcal{N}^0_{r,s})$ or $L_q(\mathcal{N}_{r,s})$ are included; other, redundant, decompositions are omitted.

Clearly $\mathcal{N}^0_{r,s} \leq \mathcal{N}^0_{t,u}$ if and only if $r \leq t$ and $s - r \mid u - t$, and in that case $L_q(\mathcal{N}^0_{r,s})$ is isomorphic to an ideal of $L_q(\mathcal{N}^0_{t,u})$. This simple observation can be useful.

For the quasivarieties $\mathcal{N}^0_{r,s}$, let us use the abbreviation $\mathbf{J}^+_{k,\ell} = \mathbf{J}_{k,\ell} + \mathbf{0}$.

We begin with the quasivariety $\mathcal{N}^0_{2,3}$, which has $s - r = 1$.
  (1) Quasicriticals
     $\mathbf{J}_{1,2}$, $\mathbf{J}_{2,3}$, $\mathbf{J}^+_{0,1}$, $\mathbf{J}^+_{1,2}$, $\mathbf{J}^+_{2,3}$.
  (2) Embeddings
     $\mathbf{J}_{1,2} \leq \mathbf{J}_{2,3}$,
     $\mathbf{J}^+_{0,1} \leq \mathbf{J}^+_{1,2} \leq \mathbf{J}^+_{2,3}$.
  (3) Subdirect decompositions
     $\mathbf{J}^+_{1,2} \leq \mathbf{J}_{1,2} \times \mathbf{J}^+_{0,1}$,
     $\mathbf{J}^+_{2,3} \leq \mathbf{J}_{2,3} \times \mathbf{J}^+_{0,1}$.

Next we consider $\mathcal{N}^0_{2,4}$, with $s - r$ prime.
  (1) Quasicriticals
     $\mathbf{J}_{1,2}$, $\mathbf{J}_{2,3}$, $\mathbf{J}^+_{0,1}$, $\mathbf{J}^+_{1,2}$, $\mathbf{J}^+_{2,3}$, $\mathbf{J}^+_{0,2}$, $\mathbf{J}^+_{1,3}$, $\mathbf{J}^+_{2,4}$.
  (2) Embeddings
     $\mathbf{J}_{1,2} \leq \mathbf{J}_{2,3}$,
     $\mathbf{J}^+_{0,1} \leq \mathbf{J}^+_{1,2} \leq \mathbf{J}^+_{2,3}$,
     $\mathbf{J}^+_{0,2} \leq \mathbf{J}^+_{1,3} \leq \mathbf{J}^+_{2,4}$.
  (3) Subdirect decompositions
     $\mathbf{J}^+_{1,2} \leq \mathbf{J}_{1,2} \times \mathbf{J}^+_{0,1}$,
     $\mathbf{J}^+_{2,3} \leq \mathbf{J}_{2,3} \times \mathbf{J}^+_{0,1}$,
     $\mathbf{J}^+_{1,3} \leq \mathbf{J}_{1,2} \times \mathbf{J}^+_{0,2}$,
     $\mathbf{J}^+_{1,3} \leq \mathbf{J}^+_{1,2} \times \mathbf{J}^+_{0,2}$,
     $\mathbf{J}^+_{2,4} \leq \mathbf{J}_{2,3} \times \mathbf{J}^+_{0,2}$,
     $\mathbf{J}^+_{2,4} \leq \mathbf{J}^+_{2,3} \times \mathbf{J}^+_{0,2}$.

The quasivariety $\mathcal{N}^0_{1,5}$ has $s - r$ being a prime power.
  (1) Quasicriticals
     $\mathbf{J}_{1,2}$, $\mathbf{J}^+_{0,1}$, $\mathbf{J}^+_{0,2}$, $\mathbf{J}^+_{0,4}$, $\mathbf{J}^+_{1,2}$, $\mathbf{J}^+_{1,3}$, $\mathbf{J}^+_{1,5}$.
  (2) Embeddings
     $\mathbf{J}^+_{0,1} \leq \mathbf{J}^+_{1,2}$,
     $\mathbf{J}^+_{0,2} \leq \mathbf{J}^+_{1,3}$,
     $\mathbf{J}^+_{0,4} \leq \mathbf{J}^+_{1,5}$.
  (3) Subdirect decompositions
     $\mathbf{J}^+_{1,2} \leq \mathbf{J}_{1,2} \times \mathbf{J}^+_{0,1}$,

$$\mathbf{J}_{1,3}^{+} \leq \mathbf{J}_{1,2} \times \mathbf{J}_{0,2}^{+},$$
$$\mathbf{J}_{1,3}^{+} \leq \mathbf{J}_{1,2}^{+} \times \mathbf{J}_{0,2}^{+},$$
$$\mathbf{J}_{1,5}^{+} \leq \mathbf{J}_{1,2} \times \mathbf{J}_{0,4}^{+},$$
$$\mathbf{J}_{1,5}^{+} \leq \mathbf{J}_{1,2}^{+} \times \mathbf{J}_{0,4}^{+},$$
$$\mathbf{J}_{1,5}^{+} \leq \mathbf{J}_{1,3}^{+} \times \mathbf{J}_{0,4}^{+}.$$

Figure 6.3 shows the lattice $L_q(\mathcal{N}_{1,3}^0)$, which is an ideal in $L_q(\mathcal{N}_{1,5}^0)$. This lattice is determined by the quasicritical algebras, embeddings, and subdirect decompositions given above, in that the quasicritical algebras give join irreducible quasivarieties, the embeddings give order relations between the join irreducibles, and the subdirect products give further order relations and join dependencies.

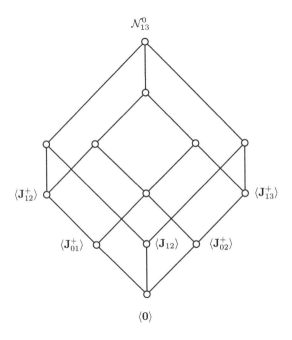

FIGURE 6.3. The lattice of subquasivarieties $L_q(\mathcal{N}_{1,3}^0)$

Finally we consider $\mathcal{N}_{1,7}^0$, an example with $s - r$ composite.

(1) Quasicriticals
   $\mathbf{J}_{1,2},\ \mathbf{J}_{0,1}^{+},\ \mathbf{J}_{0,2}^{+},\ \mathbf{J}_{0,3}^{+},\ \mathbf{J}_{0,6}^{+},\ \mathbf{J}_{1,2}^{+},\ \mathbf{J}_{1,3}^{+},\ \mathbf{J}_{1,4}^{+},\ \mathbf{J}_{1,7}^{+}.$

(2) Embeddings
   $$\mathbf{J}_{0,1}^{+} \leq \mathbf{J}_{1,2}^{+},$$
   $$\mathbf{J}_{0,2}^{+} \leq \mathbf{J}_{1,3}^{+},$$
   $$\mathbf{J}_{0,3}^{+} \leq \mathbf{J}_{1,4}^{+},$$
   $$\mathbf{J}_{0,6}^{+} \leq \mathbf{J}_{1,7}^{+}.$$

(3) Subdirect decompositions
$$\mathbf{J}_{0,6}^+ \leq \mathbf{J}_{0,2}^+ \times \mathbf{J}_{0,3}^+,$$
$$\mathbf{J}_{1,3}^+ \leq \mathbf{J}_{1,2} \times \mathbf{J}_{0,2}^+,$$
$$\mathbf{J}_{1,3}^+ \leq \mathbf{J}_{1,2}^+ \times \mathbf{J}_{0,2}^+,$$
$$\mathbf{J}_{1,4}^+ \leq \mathbf{J}_{1,2} \times \mathbf{J}_{0,3}^+,$$
$$\mathbf{J}_{1,4}^+ \leq \mathbf{J}_{1,2}^+ \times \mathbf{J}_{0,3}^+,$$
$$\mathbf{J}_{1,7}^+ \leq \mathbf{J}_{1,2} \times \mathbf{J}_{0,6}^+,$$
$$\mathbf{J}_{1,7}^+ \leq \mathbf{J}_{1,2}^+ \times \mathbf{J}_{0,6}^+,$$
$$\mathbf{J}_{1,7}^+ \leq \mathbf{J}_{1,3}^+ \times \mathbf{J}_{0,2}^+,$$
$$\mathbf{J}_{1,7}^+ \leq \mathbf{J}_{1,3}^+ \times \mathbf{J}_{0,6}^+,$$
$$\mathbf{J}_{1,7}^+ \leq \mathbf{J}_{1,4}^+ \times \mathbf{J}_{0,3}^+,$$
$$\mathbf{J}_{1,7}^+ \leq \mathbf{J}_{1,4}^+ \times \mathbf{J}_{0,6}^+.$$

6.3. Some Quasivarieties $\mathcal{N}_{r,s}$

We continue by applying Theorem 6.12 and Lemmas 6.18, 6.19, and 6.20 to various quasivarieties $\mathcal{N}_{r,s}$.

Note that $\mathcal{N}_{r,s} \leq \mathcal{N}_{t,u}$ if and only if $r \leq t$ and $s - r \mid u - t$, in which case $L_q(\mathcal{N}_{r,s})$ is isomorphic to an ideal of $L_q(\mathcal{N}_{t,u})$.

The quasivariety $\mathcal{N}_{2,3}$ has $s - r = 1$.

(1) Quasicriticals
$\mathbf{J}_{1,2}$, $\mathbf{J}_{2,3}$, $\mathbf{E} + \mathbf{E}$.

(2) Embeddings
$\mathbf{J}_{1,2} \leq \mathbf{J}_{2,3}$.

(3) Subdirect decompositions
None, they are all subdirectly irreducible.

The quasivariety $\mathcal{N}_{1,3}$ is an example with $s - r$ prime.

(1) Quasicriticals
$\mathbf{J}_{1,2}$, $\mathbf{J}_{0,2}$, $\mathbf{J}_{1,3}$, $\mathbf{E} + \mathbf{E}$, $\mathbf{J}_{0,2} + \mathbf{E}$, $\mathbf{J}_{1,3} + \mathbf{E}$.

(2) Embeddings
$$\mathbf{J}_{0,2} \leq \mathbf{J}_{1,3} \leq \mathbf{J}_{1,3} + \mathbf{E},$$
$$\mathbf{J}_{0,2} \leq \mathbf{J}_{0,2} + \mathbf{E} \leq \mathbf{J}_{1,3} + \mathbf{E}.$$

(3) Subdirect decompositions
$$\mathbf{J}_{1,3} \leq \mathbf{J}_{0,2} \times \mathbf{J}_{1,2},$$
$$\mathbf{J}_{1,3} + \mathbf{E} \leq (\mathbf{J}_{0,2} + \mathbf{E}) \times \mathbf{J}_{1,2}.$$

The quasivariety $\mathcal{N}_{1,3}$ is a convenient size to illustrate how the preceding information enables us to draw the lattice of subquasivarieties $L_q(\mathcal{N}_{1,3})$ in Figure 6.4. Again, the quasicritical algebras give join irreducible quasivarieties, the embeddings give order relations between the join irreducibles, and the subdirect products give further order relations and join dependencies.

Next, the quasivariety $\mathcal{N}_{2,4}$ also has $s - r$ prime.

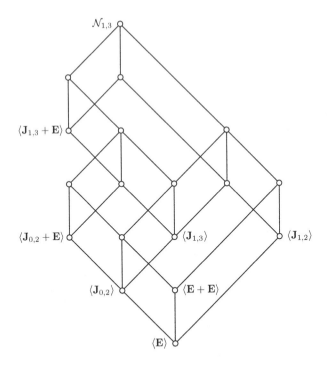

FIGURE 6.4. The lattice of subquasivarieties $L_q(\mathcal{N}_{1,3})$

(1) Quasicriticals
$\mathbf{J}_{1,2}$, $\mathbf{J}_{2,3}$, $\mathbf{J}_{0,2}$, $\mathbf{J}_{1,3}$, $\mathbf{J}_{2,4}$, $\mathbf{E} + \mathbf{E}$, $\mathbf{J}_{0,2} + \mathbf{E}$, $\mathbf{J}_{1,3} + \mathbf{E}$, $\mathbf{J}_{2,4} + \mathbf{E}$.

(2) Embeddings
$\mathbf{J}_{1,2} \leq \mathbf{J}_{2,3}$,
$\mathbf{J}_{0,2} \leq \mathbf{J}_{1,3} \leq \mathbf{J}_{2,4}$,
$\mathbf{J}_{0,2} + \mathbf{E} \leq \mathbf{J}_{1,3} + \mathbf{E} \leq \mathbf{J}_{2,4} + \mathbf{E}$,
$\mathbf{J}_{0,2} \leq \mathbf{J}_{0,2} + \mathbf{E}$,
$\mathbf{J}_{1,3} \leq \mathbf{J}_{1,3} + \mathbf{E}$,
$\mathbf{J}_{2,4} \leq \mathbf{J}_{2,4} + \mathbf{E}$.

(3) Subdirect decompositions
$\mathbf{J}_{1,3} \leq \mathbf{J}_{0,2} \times \mathbf{J}_{1,2}$,
$\mathbf{J}_{2,4} \leq \mathbf{J}_{0,2} \times \mathbf{J}_{2,3}$,
$\mathbf{J}_{1,3} + \mathbf{E} \leq (\mathbf{J}_{0,2} + \mathbf{E}) \times \mathbf{J}_{1,2}$,
$\mathbf{J}_{2,4} + \mathbf{E} \leq (\mathbf{J}_{0,2} + \mathbf{E}) \times \mathbf{J}_{2,3}$.

The quasivariety $\mathcal{N}_{1,5}$ is an example with $s - r$ a prime power.

(1) Quasicriticals
$\mathbf{J}_{1,2}$, $\mathbf{J}_{0,2}$, $\mathbf{J}_{0,4}$, $\mathbf{J}_{1,3}$, $\mathbf{J}_{1,5}$, $\mathbf{E} + \mathbf{E}$, $\mathbf{J}_{0,2} + \mathbf{J}_{0,4}$, $\mathbf{J}_{0,2} + \mathbf{J}_{1,5}$, $\mathbf{J}_{0,2} + \mathbf{E}$,
$\mathbf{J}_{0,4} + \mathbf{E}$, $\mathbf{J}_{1,3} + \mathbf{E}$, $\mathbf{J}_{1,5} + \mathbf{E}$.

(2) Embeddings

$\mathbf{J}_{0,2} \leq \mathbf{J}_{0,2} + \mathbf{J}_{0,4}$,

$\mathbf{J}_{0,2} \leq \mathbf{J}_{0,2} + \mathbf{J}_{1,5}$,

$\mathbf{J}_{0,2} \leq \mathbf{J}_{0,2} + \mathbf{E}$,

$\mathbf{J}_{0,2} \leq \mathbf{J}_{1,3} \leq \mathbf{J}_{1,3} + \mathbf{E}$,

$\mathbf{J}_{0,4} \leq \mathbf{J}_{0,2} + \mathbf{J}_{0,4}$,

$\mathbf{J}_{0,4} \leq \mathbf{J}_{0,4} + \mathbf{E}$,

$\mathbf{J}_{0,4} \leq \mathbf{J}_{1,5} \leq \mathbf{J}_{1,5} + \mathbf{E}$,

$\mathbf{J}_{1,5} \leq \mathbf{J}_{0,2} + \mathbf{J}_{1,5}$,

$\mathbf{J}_{0,2} + \mathbf{J}_{0,4} \leq \mathbf{J}_{0,2} + \mathbf{J}_{1,5}$,

$\mathbf{J}_{0,2} + \mathbf{E} \leq \mathbf{J}_{1,3} + \mathbf{E}$,

$\mathbf{J}_{0,4} + \mathbf{E} \leq \mathbf{J}_{1,5} + \mathbf{E}$.

(3) Subdirect decompositions

$\mathbf{J}_{1,3} \leq \mathbf{J}_{1,2} \times \mathbf{J}_{0,2}$,

$\mathbf{J}_{1,5} \leq \mathbf{J}_{1,2} \times \mathbf{J}_{0,4}$,

$\mathbf{J}_{0,2} + \mathbf{J}_{0,4} \leq \mathbf{J}_{0,2} \times (\mathbf{J}_{0,4} + \mathbf{E})$,

$\mathbf{J}_{0,2} + \mathbf{J}_{1,5} \leq \mathbf{J}_{0,2} \times (\mathbf{J}_{1,5} + \mathbf{E})$,

$\mathbf{J}_{1,3} + \mathbf{E} \leq (\mathbf{J}_{0,2} + \mathbf{E}) \times \mathbf{J}_{1,2}$,

$\mathbf{J}_{1,5} + \mathbf{E} \leq (\mathbf{J}_{0,4} + \mathbf{E}) \times \mathbf{J}_{1,2}$,

$\mathbf{J}_{1,5} + \mathbf{E} \leq (\mathbf{J}_{0,4} + \mathbf{E}) \times (\mathbf{J}_{1,3} + \mathbf{E})$.

Finally, the quasivariety $\mathcal{N}_{1,7}$ has $s - r$ composite.

(1) Quasicriticals

$\mathbf{J}_{1,2}$, $\mathbf{J}_{0,2}$, $\mathbf{J}_{0,3}$, $\mathbf{J}_{0,6}$, $\mathbf{J}_{1,3}$, $\mathbf{J}_{1,4}$, $\mathbf{J}_{1,7}$, $\mathbf{E} + \mathbf{E}$, $\mathbf{J}_{0,2} + \mathbf{E}$, $\mathbf{J}_{0,3} + \mathbf{E}$, $\mathbf{J}_{0,6} + \mathbf{E}$, $\mathbf{J}_{0,3} + \mathbf{J}_{0,2}$, $\mathbf{J}_{0,6} + \mathbf{J}_{0,2}$, $\mathbf{J}_{0,6} + \mathbf{J}_{0,3}$, $\mathbf{J}_{1,3} + \mathbf{E}$, $\mathbf{J}_{1,4} + \mathbf{E}$, $\mathbf{J}_{1,7} + \mathbf{E}$, $\mathbf{J}_{1,4} + \mathbf{J}_{0,2}$, $\mathbf{J}_{1,7} + \mathbf{J}_{0,2}$, $\mathbf{J}_{1,3} + \mathbf{J}_{0,3}$, $\mathbf{J}_{1,7} + \mathbf{J}_{0,3}$.

(2) Embeddings

$\mathbf{J}_{0,2} \leq \mathbf{J}_{0,2} + \mathbf{E}$,

$\mathbf{J}_{0,3} \leq \mathbf{J}_{0,3} + \mathbf{E}$,

$\mathbf{J}_{0,6} \leq \mathbf{J}_{0,6} + \mathbf{E}$,

$\mathbf{J}_{0,2} \leq \mathbf{J}_{0,3} + \mathbf{J}_{0,2}$,

$\mathbf{J}_{0,3} \leq \mathbf{J}_{0,3} + \mathbf{J}_{0,2}$,

$\mathbf{J}_{0,2} \leq \mathbf{J}_{0,6} + \mathbf{J}_{0,2}$,

$\mathbf{J}_{0,6} \leq \mathbf{J}_{0,6} + \mathbf{J}_{0,2}$,

$\mathbf{J}_{0,3} \leq \mathbf{J}_{0,6} + \mathbf{J}_{0,3}$,

$\mathbf{J}_{0,6} \leq \mathbf{J}_{0,6} + \mathbf{J}_{0,3}$,

$\mathbf{J}_{0,2} \leq \mathbf{J}_{1,3} \leq \mathbf{J}_{1,3} + \mathbf{E}$,

$\mathbf{J}_{0,3} \leq \mathbf{J}_{1,4} \leq \mathbf{J}_{1,4} + \mathbf{E}$,

$\mathbf{J}_{0,6} \leq \mathbf{J}_{1,7} \leq \mathbf{J}_{1,7} + \mathbf{E}$,

$\mathbf{J}_{0,3} + \mathbf{J}_{0,2} \leq \mathbf{J}_{1,4} + \mathbf{J}_{0,2}$,

$\mathbf{J}_{0,6} + \mathbf{J}_{0,2} \leq \mathbf{J}_{1,7} + \mathbf{J}_{0,2}$,

$\mathbf{J}_{0,2} + \mathbf{J}_{0,3} \leq \mathbf{J}_{1,3} + \mathbf{J}_{0,3}$,

$\mathbf{J}_{0,6} + \mathbf{J}_{0,3} \leq \mathbf{J}_{1,7} + \mathbf{J}_{0,3}$.

(3) Subdirect decompositions

$\mathbf{J}_{0,6} \leq \mathbf{J}_{0,2} \times \mathbf{J}_{0,3}$,

$$\mathbf{J}_{1,3} \leq \mathbf{J}_{1,2} \times \mathbf{J}_{0,2},$$
$$\mathbf{J}_{1,4} \leq \mathbf{J}_{1,2} \times \mathbf{J}_{0,3},$$
$$\mathbf{J}_{1,7} \leq \mathbf{J}_{1,2} \times \mathbf{J}_{0,6},$$
$$\mathbf{J}_{0,6} + \mathbf{E} \leq (\mathbf{J}_{0,2} + \mathbf{E}) \times (\mathbf{J}_{0,3} + \mathbf{E}),$$
$$\mathbf{J}_{1,3} + \mathbf{E} \leq \mathbf{J}_{1,2} \times (\mathbf{J}_{0,2} + \mathbf{E}),$$
$$\mathbf{J}_{1,4} + \mathbf{E} \leq \mathbf{J}_{1,2} \times (\mathbf{J}_{0,3} + \mathbf{E}),$$
$$\mathbf{J}_{1,7} + \mathbf{E} \leq \mathbf{J}_{1,2} \times (\mathbf{J}_{0,6} + \mathbf{E}),$$
$$\mathbf{J}_{1,7} + \mathbf{E} \leq (\mathbf{J}_{1,3} + \mathbf{E}) \times (\mathbf{J}_{0,3} + \mathbf{E}),$$
$$\mathbf{J}_{1,7} + \mathbf{E} \leq (\mathbf{J}_{1,4} + \mathbf{E}) \times (\mathbf{J}_{0,2} + \mathbf{E}),$$
$$\mathbf{J}_{1,7} + \mathbf{E} \leq (\mathbf{J}_{1,3} + \mathbf{E}) \times (\mathbf{J}_{0,6} + \mathbf{E}),$$
$$\mathbf{J}_{1,7} + \mathbf{E} \leq (\mathbf{J}_{1,4} + \mathbf{E}) \times (\mathbf{J}_{0,6} + \mathbf{E}),$$
$$\mathbf{J}_{0,3} + \mathbf{J}_{0,2} \leq (\mathbf{J}_{0,3} + \mathbf{E}) \times (\mathbf{J}_{0,2} + \mathbf{E}),$$
$$\mathbf{J}_{0,6} + \mathbf{J}_{0,2} \leq (\mathbf{J}_{0,6} + \mathbf{E}) \times \mathbf{J}_{0,2},$$
$$\mathbf{J}_{0,6} + \mathbf{J}_{0,2} \leq (\mathbf{J}_{0,3} + \mathbf{E}) \times \mathbf{J}_{0,2},$$
$$\mathbf{J}_{0,6} + \mathbf{J}_{0,3} \leq (\mathbf{J}_{0,6} + \mathbf{E}) \times \mathbf{J}_{0,3},$$
$$\mathbf{J}_{0,6} + \mathbf{J}_{0,3} \leq (\mathbf{J}_{0,2} + \mathbf{E}) \times \mathbf{J}_{0,3},$$
$$\mathbf{J}_{1,4} + \mathbf{J}_{0,2} \leq \mathbf{J}_{1,2} \times (\mathbf{J}_{0,3} + \mathbf{J}_{0,2}),$$
$$\mathbf{J}_{1,4} + \mathbf{J}_{0,2} \leq (\mathbf{J}_{1,4} + \mathbf{E}) \times (\mathbf{J}_{0,2} + \mathbf{E}),$$
$$\mathbf{J}_{1,3} + \mathbf{J}_{0,2} \leq \mathbf{J}_{1,2} \times (\mathbf{J}_{0,2} + \mathbf{J}_{0,3}),$$
$$\mathbf{J}_{1,3} + \mathbf{J}_{0,3} \leq (\mathbf{J}_{1,3} + \mathbf{E}) \times (\mathbf{J}_{0,3} + \mathbf{E}),$$
$$\mathbf{J}_{1,4} + \mathbf{J}_{0,2} \leq \mathbf{J}_{1,2} \times (\mathbf{J}_{0,3} + \mathbf{J}_{0,2}),$$
$$\mathbf{J}_{1,7} + \mathbf{J}_{0,2} \leq \mathbf{J}_{1,2} \times (\mathbf{J}_{0,6} + \mathbf{J}_{0,2}),$$
$$\mathbf{J}_{1,7} + \mathbf{J}_{0,2} \leq (\mathbf{J}_{1,4} + \mathbf{E}) \times \mathbf{J}_{0,2},$$
$$\mathbf{J}_{1,7} + \mathbf{J}_{0,3} \leq \mathbf{J}_{1,2} \times (\mathbf{J}_{0,6} + \mathbf{J}_{0,3}),$$
$$\mathbf{J}_{1,7} + \mathbf{J}_{0,3} \leq (\mathbf{J}_{1,7} + \mathbf{E}) \times \mathbf{J}_{0,3},$$
$$\mathbf{J}_{1,7} + \mathbf{J}_{0,3} \leq (\mathbf{J}_{1,3} + \mathbf{E}) \times \mathbf{J}_{0,3}.$$

To see examples of quasicritical algebras with more than two components, consider $\mathcal{N}_{0,12}$. These include $\mathbf{J}_{0,2}$, $\mathbf{J}_{0,3}$, $\mathbf{J}_{0,4}$, $\mathbf{J}_{0,6}$, $\mathbf{J}_{0,12}$, and $\mathbf{E} + \mathbf{E}$. Also, $\mathbf{A} + \mathbf{B}$ is quasicritical for any distinct pair of $\mathbf{E}$, $\mathbf{J}_{0,2}$, $\mathbf{J}_{0,3}$, $\mathbf{J}_{0,4}$, $\mathbf{J}_{0,6}$, and $\mathbf{J}_{0,12}$. Finally, the triples $\mathbf{J}_{0,2} + \mathbf{J}_{0,3} + \mathbf{J}_{0,4}$, $\mathbf{J}_{0,2} + \mathbf{J}_{0,3} + \mathbf{J}_{0,12}$ and $\mathbf{J}_{0,3} + \mathbf{J}_{0,4} + \mathbf{J}_{0,6}$ are quasicritical by Theorem 6.12, for a total of 24 algebras.

It is straightforward to do the corresponding analysis for varieties such as $\mathcal{N}_{0,30}$ where $s - r$ has more than two distinct prime factors, but this would strain even the authors' capacity for tedium.

CHAPTER 7

Pure Unary Relational Structures

7.1. Pure Unary Relational Quasivarieties

The methods developed for quasivarieties in the preceding sections apply to systems more general than just algebras. For something a little different, let us turn our attention to *pure unary relational structures*, which are sets with finitely many unary predicates $A_1, \ldots, A_k$.

Thus the language we are working in has equality and these predicates. A structure is defined by $\mathbf{S} = \langle S, A_1^{\mathbf{S}}, \ldots, A_k^{\mathbf{S}} \rangle$ where each $A_j^{\mathbf{S}} \subseteq S$ denotes those $x \in S$ for which the predicate $A_j x$ holds in $\mathbf{S}$. A homomorphism $h : \mathbf{S} \to \mathbf{T}$ is a map $h : S \to T$ such that $h(A_j^{\mathbf{S}}) \subseteq A_j^{\mathbf{T}}$ for each j. The *kernel* of a homomorphism h is a $(k+1)$-tuple $\ker h = \langle \eta_0, \eta_1, \ldots, \eta_k \rangle$ where

- $\eta_0 = \{(x, y) \in S^2 : h(x) = h(y)\}$,
- $\eta_j = \{x \in S : h(x) \in A_j^{\mathbf{T}}\}$ for $1 \leq j \leq k$.

In other words, $\eta_0 = h^{-1}(=^{\mathbf{T}})$ and $\eta_j = h^{-1}(A_j^{\mathbf{T}})$ for $1 \leq j \leq k$.

Correspondingly, a *congruence* on a relational structure is a $(k+1)$-tuple $\varphi = \langle \varphi_0, \varphi_1, \ldots, \varphi_k \rangle$ such that

- φ_0 is an equivalence relation on S,
- $A_j^{\mathbf{S}} \subseteq \varphi_j \subseteq S$ for $1 \leq j \leq k$,
- $(x, y) \in \varphi_0$ and $x \in \varphi_j$ implies $y \in \varphi_j$ for $1 \leq j \leq k$.

Thus the kernel of a homomorphism is a congruence. We can use congruences to form factor structures, and the Isomorphism Theorems hold.

Congruences are naturally ordered by componentwise inclusion, so that $\varphi \leq \psi$ iff $\varphi_j \subseteq \psi_j$ for $0 \leq j \leq k$. The congruences on a structure $\mathbf{S}$ form an algebraic lattice Con $\mathbf{S}$, with least element denoted Δ. Note that $\Delta = \langle =^{\mathbf{S}}, A_1^{\mathbf{S}}, \ldots, A_k^{\mathbf{S}} \rangle$.

For a relational structure $\mathbf{S}$ and $x \in S$, let $\rho(x)$ be the set of predicates A_j such that $A_j x$ holds in $\mathbf{S}$, i.e., such that $x \in A_j^{\mathbf{S}}$.

We use the suggestive notation $[x \equiv y]$ for the congruence $\mathrm{Cg}(x, y)$, and $[Ax]$ for the congruence that adds the predicate Ax to those of $\mathbf{S}$. Note that the congruence $[x \equiv y]$ includes the pair (x, y) in its first component and the predicates Ax and Ay for every $A \in \rho(x) \cup \rho(y)$.

An *embedding* of a structure $\mathbf{S}$ into a structure $\mathbf{T}$ is a one-to-one homomorphism $h : \mathbf{S} \to \mathbf{T}$ such that for every predicate A_j and every $x \in S$, we have $h(x) \in A_j^{\mathbf{T}}$ if and only if $x \in A_j^{\mathbf{S}}$. Thus $\mathbf{S}$ embeds into $\mathbf{T}$ if and only if $\mathbf{S}$ is isomorphic to a subset of T with the induced predicates. As usual, we write $\mathbf{S} \leq \mathbf{T}$ to denote the existence of an embedding.

Products and subdirect products behave as usual. Note that the empty product is a 1-element structure with all predicates holding.

Quasivarieties are classes closed under embedding, direct products, and ultraproducts. Again they are the models of a set of Horn sentences in the language. For a quasivariety $\mathcal{K}$ and a structure $\mathbf{S}$, the congruences $\varphi \in \text{Con } \mathbf{S}$ such that $\mathbf{S}/\varphi \in \mathcal{K}$ are closed under intersection, and form an algebraic lattice $\text{Con}_{\mathcal{K}} \mathbf{S}$.

A finite structure is *quasicritical* if it is not in the quasivariety generated by its proper substructures.

There is one minor adjustment that must be made before applying results for finite algebras to relational structures. Some kinds of theorems for algebras are normally proved by induction on the size of the algebra. In each such case, the proof can be modified to apply to more general structures by using induction on depth in the congruence lattice. This is a straightforward exercise, which we have done for the results in this monograph.

Recall that we refer to the quasi-equations $\varepsilon_{\mathbf{T},\alpha}$ as the *semi-splitting quasi-equations* of a structure $\mathbf{T}$, as their satisfaction by a quasivariety corresponds to omitting the structures in $\mathbf{N}(\mathbf{T}, \alpha)$ (Theorem 3.4). In this context, when $\varepsilon = \varepsilon_{\mathbf{T},\alpha}$ and $\mathbf{N}(\mathbf{T}, \alpha) = \{\mathbf{R}_1, \dots, \mathbf{R}_k\}$, we may say that ε *is characterized by omitting* $\{\mathbf{R}_1, \dots, \mathbf{R}_k\}$.

7.2. Quasicritical Lemmas

The next few lemmas, giving necessary conditions for a pure unary structure to be quasicritical, are variations on a common theme.

LEMMA 7.1. *If a pure unary relational structure* $\mathbf{S}$ *contains elements* x, y, z, t *with* $x \neq y$, $z \neq t$, $y \neq t$ *and* $\{x, y\} \neq \{z, t\}$ *such that* $\rho(x) \supseteq \rho(y)$ *and* $\rho(z) \supseteq \rho(t)$, *then* $\mathbf{S}$ *is not quasicritical.*

PROOF. Check that $\mathbf{S}/[x \equiv y] \cong \mathbf{S} - \{y\} \leq \mathbf{S}$ and $\mathbf{S}/[z \equiv t] \cong \mathbf{S} - \{t\} \leq \mathbf{S}$ while $[x \equiv y] \wedge [z \equiv t] = \Delta$. □

LEMMA 7.2. *If a pure unary relational structure* $\mathbf{S}$ *contains distinct elements* x, y, t *such that* $\rho(x) \supseteq \rho(y) \cup \rho(t)$, *then* $\mathbf{S}$ *is not quasicritical.*

PROOF. Apply Lemma 7.1 with $z = x$. □

LEMMA 7.3. *If a pure unary relational structure* $\mathbf{S}$ *contains distinct elements* x, y, z *such that* $\rho(x) = \rho(y) \cap \rho(z)$, *then* $\mathbf{S}$ *is not quasicritical.*

PROOF. Check that $\mathbf{S}/[x \equiv y] \cong \mathbf{S} - \{x\} \leq \mathbf{S}$ and $\mathbf{S}/[x \equiv z] \cong \mathbf{S} - \{x\} \leq \mathbf{S}$ and $[x \equiv y] \wedge [x \equiv z] = \Delta$. □

These lemmas bound the size of a quasicritical pure unary relational structure. We use the following combinatorial argument.

LEMMA 7.4. *Let* $\mathbf{P}$ *be a finite ordered set of width* w, *and let* S *be a set and* $\sigma : S \to \mathbf{P}$ *any map. If* $|S| > w + 1$, *then there exist elements* x, y, z, $t \in S$ *with* $x \neq y$, $z \neq t$, $y \neq t$ *and* $\{x,y\} \neq \{z,t\}$ *such that* $\sigma(x) \geq \sigma(y)$ *and* $\sigma(z) \geq \sigma(t)$.

PROOF. If there exist distinct elements x, y, $t \in S$ with $\sigma(x) \geq \sigma(y)$ and $\sigma(x) \geq \sigma(t)$, then the conclusion holds with $z = x$. To simplify matters, assume this does not occur.

Let $\mathbf{P}'$ be the image $\sigma(\mathbf{P})$, and let A be a maximal-sized antichain in $\mathbf{P}'$ that is maximal in the natural order on maximal-sized antichains; see [52] or [67]. Then $|A| \leq w$, and we can choose a subset $X \subseteq S$ such that $\sigma|_X$ is a bijection onto A. Since $|S| > |X| + 1$, we can choose two distinct elements y, $t \in S \setminus X$.

Because A is a maximal antichain in $\mathbf{P}'$, there are elements x_1, $x_2 \in X$ such that $\sigma(y)$ is comparable to $\sigma(x_1)$ and $\sigma(t)$ is comparable to $\sigma(x_2)$. If $\sigma(y) \leq \sigma(x_1)$ and $\sigma(t) \leq \sigma(x_2)$, we get the desired conclusion, regardless of whether $x_1 = x_2$. If say $\sigma(y) \geq \sigma(x_1)$ and $\sigma(t) \leq \sigma(x_2)$, we obtain the conclusion by interchanging the roles of y and x_1.

By symmetry, that leaves the case $\sigma(y) > \sigma(x_1)$ and $\sigma(t) > \sigma(x_2)$. In view of our initial assumption, $\sigma(y) \not\geq \sigma(x_i)$ for $i \neq 1$, and $\sigma(t) \not\geq \sigma(x_j)$ for $j \neq 2$. This implies that $\sigma(y)$ and $\sigma(t)$ are incomparable. If $x_1 \neq x_2$, we can replace $\sigma(x_1)$ and $\sigma(x_2)$ by $\sigma(y)$ and $\sigma(t)$, respectively, to obtain a maximal-sized antichain that is strictly above A in the order on maximal-sized antichains of $\mathbf{P}'$, contrary to the maximality of A. On the other hand, if $x_1 = x_2$, the antichain $A \setminus \{\sigma(x_1)\} \cup \{\sigma(y), \sigma(t)\}$ is an antichain of strictly larger size, violating that assumption. $\square$

Let us apply Lemma 7.4 to a pure unary relational structure $\mathbf{S}$ in a type with k predicates, with $\mathbf{P}$ being the lattice of subsets of the relations $A_1, \ldots, A_k$ and σ being the map ρ. Since the lattice of subsets of a k-element set has width $\binom{k}{\lfloor \frac{k}{2} \rfloor}$ by Sperner's Theorem [162], we obtain our desired bound from Lemma 7.1.

THEOREM 7.5. *If the type of a pure unary relational structure* $\mathbf{S}$ *has* k *unary relations and* $|S| > \binom{k}{\lfloor \frac{k}{2} \rfloor} + 1$, *then* $\mathbf{S}$ *is not quasicritical.*

Let $\mathcal{R}_k$ denote the quasivariety of all pure unary relational structures with k predicates.

COROLLARY 7.6. *For each finite* $k \geq 0$, *the lattice* $\mathrm{L}_q(\mathcal{R}_k)$ *is finite.*

For comparison, we note that the quasivariety of structures with one binary relation, i.e., graphs, is Q-universal (Adams and Dziobiak [6]; see also [40], [110], [161]).

7.3. One Unary Relation

The quasivariety $\mathcal{R}_1$ consists of structures with a single unary predicate A. It follows from Theorem 7.5 that a quasicritical structure in $\mathcal{R}_1$ has at most 2 elements. Let us determine the quasicritical structures, and then the lattice of subquasivarieties $L_q(\mathcal{R}_1)$.

The quasicritical structures in $\mathcal{R}_1$ are the structures $\mathbf{H}_i$ $(1 \leq i \leq 4)$ described in Table 7.1. The embedding relations between quasicriticals are $\mathbf{H}_1 \leq \mathbf{H}_3$ and $\mathbf{H}_1 \leq \mathbf{H}_4$. As an aid to the calculations below, the congruence lattice of $\mathbf{H}_4$, the free $\mathcal{R}_1$-structure on 2 elements, is drawn in Figure 7.1.

Structure	Elements	$A^{\mathbf{H}_i}$
$\mathbf{H}_0$	x	x
$\mathbf{H}_1$	x	$\varnothing$
$\mathbf{H}_2$	x, y	x, y
$\mathbf{H}_3$	x, y	x
$\mathbf{H}_4$	x, y	$\varnothing$

TABLE 7.1. Quasicritical structures in $\mathcal{R}_1$, and the trivial structure $\mathbf{H}_0$

Let $\mathbf{H}_0$ denote $\{x\}$ with the predicate Ax. This is the trivial structure and not quasicritical; it is in every subquasivariety of $\mathcal{R}_1$ as the empty product.

The 1-generated free structure $\mathbf{H}_1$ is $\{x\}$ with the relation A empty. As the structure is simple, $\langle \mathbf{H}_1 \rangle$ is join prime. Its splitting quasi-equation is Ax, so that $\mathbf{S} \in \langle Ax \rangle$ if and only if $\mathbf{H}_1 \not\leq \mathbf{S}$.

The structure $\mathbf{H}_2$ is $\{x, y\}$ with Ax, Ay. This is also simple and so $\langle \mathbf{H}_2 \rangle$ is join prime. Its splitting quasi-equation is $Ax \,\&\, Ay \to x \approx y$, characterized by omitting $\mathbf{H}_2$.

The structure $\mathbf{H}_3$ is $\{x, y\}$ with Ax. Note $\mathbf{H}_1 \leq \mathbf{H}_3$. This has a 3-element chain for its congruence lattice, with atom $[Ay]$. Its splitting quasi-equation is $Ax \to Ay$, characterized by omitting $\mathbf{H}_3$.

The 2-generated free structure $\mathbf{H}_4$ is $\{x, y\}$ with the relation A empty. Note $\mathbf{H}_1 \leq \mathbf{H}_4$ and $\mathbf{H}_4 \leq \mathbf{H}_1 \times \mathbf{H}_2$ and $\mathbf{H}_4 \leq \mathbf{H}_1 \times \mathbf{H}_3 \leq \mathbf{H}_3^2$. Its 6-element congruence lattice has atoms $[Ax]$, $[Ay]$ and $[x \equiv y]$. The semi-splitting quasi-equations are $x \approx y$ (omitting $\mathbf{H}_2, \mathbf{H}_3, \mathbf{H}_4$) and Ax (omitting $\mathbf{H}_1$).

PROPOSITION 7.7. *The quasicritical structures in $\mathcal{R}_1$ are $\mathbf{H}_1$, $\mathbf{H}_2$, $\mathbf{H}_3$, and $\mathbf{H}_4$.*

Combining this information enables us to draw the lattice $L_q(\mathcal{R}_1)$ of Figure 7.2. It is the finite lattice with join irreducibles $\langle \mathbf{H}_i \rangle$ for $1 \leq i \leq 4$ satisfying $\langle \mathbf{H}_1 \rangle \leq \langle \mathbf{H}_4 \rangle \leq \langle \mathbf{H}_3 \rangle$ and $\langle \mathbf{H}_4 \rangle \leq \langle \mathbf{H}_1 \rangle \vee \langle \mathbf{H}_2 \rangle$.

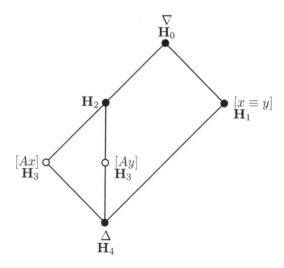

FIGURE 7.1. Con $\mathbf{H}_4$ where $\mathbf{H}_4 = \mathcal{F}_2(\mathcal{R}_1)$

7.4. Two Unary Relations

Now we can do the same type of analysis for the class $\mathcal{R}_2$ of structures with two unary relations, say A and B. By Theorem 7.5, quasicritical structures in $\mathcal{R}_2$ have at most 3 elements. The quasicritical structures in $\mathcal{R}_2$ are the structures $\mathbf{T}_i$ ($1 \leq i \leq 3$) and $\mathbf{U}_j$ ($1 \leq j \leq 12$) described in Table 7.2. The embeddings of these quasicritical structures are shown in Figure 7.3.

Let $\mathbf{T}_0$ denote $\{x\}$ with the predicates Ax, Bx. This is the trivial structure and not quasicritical.

The structure $\mathbf{T}_1$ is $\{x\}$ with Ax. This structure is simple, and $\langle \mathbf{T}_1 \rangle$ is join prime. Its splitting quasi-equation is $Ax \to Bx$, characterized by omitting $\mathbf{T}_1$.

Symmetrically, $\mathbf{T}_2$ is $\{x\}$ with Bx. The structure is simple, and $\langle \mathbf{T}_2 \rangle$ is join prime. Its splitting quasi-equation is $Bx \to Ax$, characterized by omitting $\mathbf{T}_2$.

The 1-generated free structure $\mathbf{T}_3$ is $\{x\}$ with both relations A, B empty. Note that $\mathbf{T}_3 \leq \mathbf{T}_1 \times \mathbf{T}_2$. Its congruence lattice has two atoms, $[Ax]$ and $[Bx]$. The semi-splitting quasi-equations are Ax, characterized by omitting $\mathbf{T}_2$, $\mathbf{T}_3$, and Bx, characterized by omitting $\mathbf{T}_1$, $\mathbf{T}_3$.

The structure $\mathbf{U}_1$ is $\{x,y\}$ with Ax, Bx, Ay, By. This structure is simple, so $\langle \mathbf{U}_1 \rangle$ is join prime. Its splitting quasi-equation is given by $Ax \,\&\, Bx \,\&\, Ay \,\&\, By \to x \approx y$, characterized by omitting $\mathbf{U}_1$.

The structure $\mathbf{U}_2$ is $\{x,y\}$ with Ax, Bx, Ay. Note $\mathbf{T}_1 \leq \mathbf{U}_2$. Its congruence lattice is a 3-element chain with atom $[By]$. Thus $\langle \mathbf{U}_2 \rangle$ is join prime with splitting quasi-equation $Ax \,\&\, Bx \,\&\, Ay \to By$, characterized by omitting $\mathbf{U}_2$.

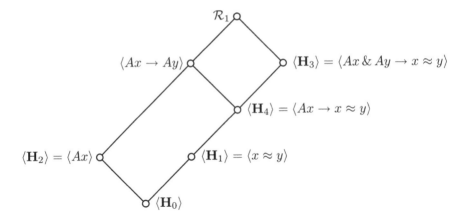

FIGURE 7.2. Quasivariety lattice $L_q(\mathcal{R}_1)$

structure	elements	$A^{\mathbf{U}_i}$	$B^{\mathbf{U}_i}$
$\mathbf{T}_0$	x	x	x
$\mathbf{T}_1$	x	x	$\varnothing$
$\mathbf{T}_2$	x	$\varnothing$	x
$\mathbf{T}_3$	x	$\varnothing$	$\varnothing$
$\mathbf{U}_1$	x, y	x, y	x, y
$\mathbf{U}_2$	x, y	x, y	x
$\mathbf{U}_3$	x, y	x, y	$\varnothing$
$\mathbf{U}_4$	x, y	x	x, y
$\mathbf{U}_5$	x, y	x	x
$\mathbf{U}_6$	x, y	x	y
$\mathbf{U}_7$	x, y	x	$\varnothing$
$\mathbf{U}_8$	x, y	$\varnothing$	x, y
$\mathbf{U}_9$	x, y	$\varnothing$	x
$\mathbf{U}_{10}$	x, y	$\varnothing$	$\varnothing$
$\mathbf{U}_{11}$	x, y, z	x, y	z
$\mathbf{U}_{12}$	x, y, z	x	y, z

TABLE 7.2. Quasicritical structures in $\mathcal{R}_2$

The structure $\mathbf{U}_3$ is $\{x, y\}$ with Ax, Ay. Note that $\mathbf{T}_1 \leq \mathbf{U}_3 \leq \mathbf{T}_1 \times \mathbf{U}_2 \leq \mathbf{U}_2^2$ and $\mathbf{U}_3 \leq \mathbf{T}_1 \times \mathbf{U}_1$. The congruence lattice has three atoms, $[Bx]$, $[By]$ and $[x \equiv y]$, giving two minimal reflection congruences. The semi-splitting quasi-equations are $Ax \& Ay \to Bx$, equivalent to $Ax \to Bx$ and characterized by omitting $\mathbf{T}_1$, and $Ax \& Ay \to x \approx y$, characterized by omitting $\mathbf{U}_1$, $\mathbf{U}_2$, $\mathbf{U}_3$.

Symmetric to $\mathbf{U}_2$, the structure $\mathbf{U}_4$ is $\{x, y\}$ with Ax, Bx, By. Note $\mathbf{T}_2 \leq \mathbf{U}_4$. Its congruence lattice is a 3-element chain with atom $[Ay]$. Thus $\langle \mathbf{U}_4 \rangle$ is

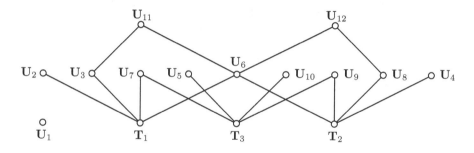

FIGURE 7.3. Quasicritical embeddings in $\mathcal{R}_2$

join prime with splitting quasi-equation $Ax \,\&\, Bx \,\&\, By \to Ay$, characterized by omitting $\mathbf{U}_4$.

The structure $\mathbf{U}_5$ is $\{x, y\}$ with Ax, Bx. Note that $\mathbf{T}_3 \leq \mathbf{U}_5 \leq \mathbf{U}_2 \times \mathbf{U}_4$. The congruence lattice has two atoms, $[Ay]$ and $[By]$. The semi-splitting quasi-equations are $Ax \,\&\, Bx \to Ay$, characterized by omitting $\mathbf{U}_4$ and $\mathbf{U}_5$, and $Ax \,\&\, Bx \to By$, characterized by omitting $\mathbf{U}_2$ and $\mathbf{U}_5$.

The structure $\mathbf{U}_6$ is $\{x, y\}$ with Ax, By. Note that $\mathbf{T}_1$, $\mathbf{T}_2 \leq \mathbf{U}_6 \leq \mathbf{U}_2 \times \mathbf{U}_4$. Its congruence lattice has two atoms, $[Ay]$ and $[Bx]$. The semi-splitting quasi-equations are $Ax \,\&\, By \to Ay$, characterized by omitting $\mathbf{U}_4$ and $\mathbf{U}_6$, and $Ax \,\&\, By \to Bx$, characterized by omitting $\mathbf{U}_2$ and $\mathbf{U}_6$.

The structure $\mathbf{U}_7$ is $\{x, y\}$ with Ax. Its congruence lattice is drawn in Figure 7.4. Note that $\mathbf{T}_1$, $\mathbf{T}_3 \leq \mathbf{U}_7 \leq \mathbf{T}_1 \times \mathbf{U}_6 \leq \mathbf{U}_6^2$. Also $\mathbf{U}_7 \leq \mathbf{T}_1 \times \mathbf{U}_4$ and $\mathbf{U}_7 \leq \mathbf{T}_1 \times \mathbf{U}_5$. Its congruence lattice has three atoms, $[Bx]$, $[By]$ and $[Ay]$. The semi-splitting quasi-equations are $Ax \to Bx$, characterized by omitting $\mathbf{T}_1$; $Ax \to By$, characterized by omitting $\mathbf{T}_1$ and $\mathbf{U}_5$; and $Ax \to Ay$, characterized by omitting $\mathbf{U}_4$, $\mathbf{U}_5$, $\mathbf{U}_6$, and $\mathbf{U}_7$. Observe that $\langle Ax \to By \rangle \leq \langle Ax \to Bx \rangle$, so that the quasivariety $\langle Ax \to By \rangle$ is not in $\kappa(\mathbf{U}_7)$, and indeed the reflective closure $\gamma^{\mathbf{U}_7}([By])$ is above $[Bx]$.

Symmetric to $\mathbf{U}_3$, the structure $\mathbf{U}_8$ is $\{x, y\}$ with Bx, By. Note that $\mathbf{T}_2 \leq \mathbf{U}_8 \leq \mathbf{T}_2 \times \mathbf{U}_4 \leq \mathbf{U}_4^2$ and $\mathbf{U}_8 \leq \mathbf{T}_2 \times \mathbf{U}_1$. The congruence lattice has three atoms, $[Ax]$, $[Ay]$ and $[x \equiv y]$, giving two minimal reflection congruences. The semi-splitting quasi-equations are $Bx \,\&\, By \to Ax$, equivalent to $Bx \to Ax$ and characterized by omitting $\mathbf{T}_2$, and $Bx \,\&\, By \to x \approx y$, characterized by omitting $\mathbf{U}_1$, $\mathbf{U}_4$, $\mathbf{U}_8$.

Symmetric to $\mathbf{U}_7$, the structure $\mathbf{U}_9$ is $\{x, y\}$ with Bx. Note that $\mathbf{T}_2$, $\mathbf{T}_3 \leq \mathbf{U}_9 \leq \mathbf{T}_2 \times \mathbf{U}_6 \leq \mathbf{U}_6^2$. Also $\mathbf{U}_9 \leq \mathbf{T}_2 \times \mathbf{U}_2$ and $\mathbf{U}_9 \leq \mathbf{T}_2 \times \mathbf{U}_5$. Its congruence lattice has three atoms, $[Ax]$, $[Ay]$, and $[By]$. The semi-splitting quasi-equations for $\mathbf{U}_9$ are $Bx \to Ax$, characterized by omitting $\mathbf{T}_2$; $Bx \to Ay$, characterized by omitting $\mathbf{T}_2$ and $\mathbf{U}_5$; and $Bx \to By$, characterized by omitting $\mathbf{U}_2$, $\mathbf{U}_5$, $\mathbf{U}_6$, and $\mathbf{U}_9$. Similar to the situation with $\mathbf{U}_7$, the quasivariety $\langle Bx \to Ay \rangle$ is not in $\kappa(\mathbf{U}_9)$.

The free structure $\mathbf{U}_{10}$ is $\{x, y\}$ with both relations A, B empty. Of course $\mathbf{T}_3 \leq \mathbf{U}_{10}$. Its subdirect decompositions include $\mathbf{U}_{10} \leq \mathbf{T}_3 \times \mathbf{U}_j$

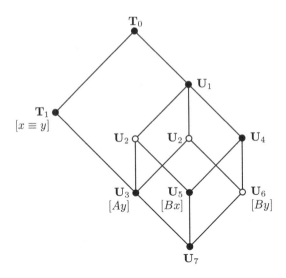

FIGURE 7.4. Con $\mathbf{U}_7$ where $\mathbf{U}_7$ is $\{x, y\}$ with the predicate Ax. Solid dots indicate reflection congruences.

for $1 \leq j \leq 9$, $\mathbf{U}_{10} \leq \mathbf{T}_1 \times \mathbf{U}_8$, $\mathbf{U}_{10} \leq \mathbf{T}_1 \times \mathbf{U}_9$, $\mathbf{U}_{10} \leq \mathbf{T}_2 \times \mathbf{U}_3$, and $\mathbf{U}_{10} \leq \mathbf{T}_2 \times \mathbf{U}_7$. Some of these give inclusions for quasivarieties, while others are not minimal covers in $\mathsf{L}_q(\mathcal{R}_2)$. In particular, $\mathbf{U}_{10} \leq \mathbf{T}_3 \times \mathbf{U}_5 \leq \mathbf{U}_5^2$, so $\langle \mathbf{U}_{10} \rangle \leq \langle \mathbf{U}_5 \rangle$. Similarly, $\langle \mathbf{U}_{10} \rangle \leq \langle \mathbf{U}_7 \rangle \leq \langle \mathbf{U}_6 \rangle$ and $\langle \mathbf{U}_{10} \rangle \leq \langle \mathbf{U}_9 \rangle \leq \langle \mathbf{U}_6 \rangle$. Since $\langle \mathbf{U}_3 \rangle < \langle \mathbf{U}_2 \rangle$, the join cover $\langle \mathbf{U}_{10} \rangle \leq \langle \mathbf{T}_3 \rangle \vee \langle \mathbf{U}_2 \rangle$ is not minimal, nor is the join cover $\langle \mathbf{U}_{10} \rangle \leq \langle \mathbf{T}_3 \rangle \vee \langle \mathbf{U}_4 \rangle$. The remaining decompositions figure into our description of $\mathsf{L}_q(\mathcal{R}_2)$, given after Proposition 7.8. The congruence lattice of $\mathbf{U}_{10}$ has five atoms, which give the three semi-splitting equations Ax, characterized by omitting $\mathbf{T}_2$ and $\mathbf{T}_3$; Bx, characterized by omitting $\mathbf{T}_1$ and $\mathbf{T}_3$; and $x \approx y$, characterized by omitting $\mathbf{U}_j$ for $1 \leq j \leq 10$.

The structure $\mathbf{U}_{11}$ is $\{x, y, z\}$ with Ax, Ay, Bz. Note that $\mathbf{U}_3, \mathbf{U}_6 \leq \mathbf{U}_{11}$. Its subdirect decompositions include $\mathbf{U}_{11} \leq \mathbf{U}_1 \times \mathbf{U}_6$ and $\mathbf{U}_{11} \leq \mathbf{U}_2 \times \mathbf{U}_6 \leq \mathbf{U}_2^2 \times \mathbf{U}_4$. The congruence lattice has four atoms, giving the three semi-splitting quasi-equations $Ax \,\&\, Ay \,\&\, Bz \rightarrow x \approx y$, characterized by omitting $\mathbf{U}_1$, $\mathbf{U}_2$, $\mathbf{U}_{11}$; $Ax \,\&\, Ay \,\&\, Bz \rightarrow Bx$, which is equivalent to $Ax \,\&\, Bz \rightarrow Bx$, characterized by omitting $\mathbf{U}_2$, $\mathbf{U}_6$; $Ax \,\&\, Ay \,\&\, Bz \rightarrow Az$, which is equivalent to $Ax \,\&\, Bz \rightarrow Az$, characterized by omitting $\mathbf{U}_4$, $\mathbf{U}_6$.

Symmetric to $\mathbf{U}_{11}$, the structure $\mathbf{U}_{12}$ is $\{x, y, z\}$ with Ax, By, Bz. Note that $\mathbf{U}_8, \mathbf{U}_6 \leq \mathbf{U}_{12}$. Its subdirect decompositions include $\mathbf{U}_{12} \leq \mathbf{U}_1 \times \mathbf{U}_6$ and $\mathbf{U}_{12} \leq \mathbf{U}_4 \times \mathbf{U}_6 \leq \mathbf{U}_2 \times \mathbf{U}_4^2$. There are four atoms, giving the three semi-splitting quasi-equations $Ax \,\&\, By \,\&\, Bz \rightarrow y \approx z$, characterized by omitting $\mathbf{U}_1$, $\mathbf{U}_4$, $\mathbf{U}_{12}$; $Ax \,\&\, By \,\&\, Bz \rightarrow Az$, which is equivalent to $Ax \,\&\, Bz \rightarrow Az$, characterized by omitting $\mathbf{U}_4$, $\mathbf{U}_6$; $Ax \,\&\, By \,\&\, Bz \rightarrow Bx$, which is equivalent to $Ax \,\&\, Bz \rightarrow Bx$, characterized by omitting $\mathbf{U}_2$, $\mathbf{U}_6$.

PROPOSITION 7.8. *The quasicritical structures in $\mathcal{R}_2$ are* $\mathbf{T}_1$, $\mathbf{T}_2$, $\mathbf{T}_3$ *and* $\mathbf{U}_j$ *for* $1 \leq j \leq 12$.

Thus $L_q(\mathcal{R}_2)$ is the finite lattice with join irreducibles $\langle \mathbf{T}_i \rangle$ for $1 \leq i \leq 3$ and $\langle \mathbf{U}_j \rangle$ for $1 \leq j \leq 12$ satisfying

(1) $\langle \mathbf{T}_3 \rangle \leq \langle \mathbf{T}_1 \rangle \vee \langle \mathbf{T}_2 \rangle$,

(2) $\langle \mathbf{T}_1 \rangle \leq \langle \mathbf{U}_3 \rangle \leq \langle \mathbf{U}_2 \rangle$,

(3) $\langle \mathbf{U}_3 \rangle \leq \langle \mathbf{T}_1 \rangle \vee \langle \mathbf{U}_1 \rangle$,

(4) $\langle \mathbf{U}_5 \rangle \leq \langle \mathbf{U}_2 \rangle \vee \langle \mathbf{U}_4 \rangle$,

(5) $\langle \mathbf{U}_6 \rangle \leq \langle \mathbf{U}_2 \rangle \vee \langle \mathbf{U}_4 \rangle$,

(6) $\langle \mathbf{T}_1 \rangle \leq \langle \mathbf{U}_7 \rangle$,

(7) $\langle \mathbf{U}_7 \rangle \leq \langle \mathbf{U}_6 \rangle$,

(8) $\langle \mathbf{U}_7 \rangle \leq \langle \mathbf{T}_1 \rangle \vee \langle \mathbf{U}_4 \rangle$,

(9) $\langle \mathbf{U}_7 \rangle \leq \langle \mathbf{T}_1 \rangle \vee \langle \mathbf{U}_5 \rangle$,

(10) $\langle \mathbf{T}_2 \rangle \leq \langle \mathbf{U}_8 \rangle \leq \langle \mathbf{U}_4 \rangle$,

(11) $\langle \mathbf{U}_8 \rangle \leq \langle \mathbf{T}_2 \rangle \vee \langle \mathbf{U}_1 \rangle$,

(12) $\langle \mathbf{T}_2 \rangle \leq \langle \mathbf{U}_9 \rangle$,

(13) $\langle \mathbf{U}_9 \rangle \leq \langle \mathbf{U}_6 \rangle$,

(14) $\langle \mathbf{U}_9 \rangle \leq \langle \mathbf{T}_2 \rangle \vee \langle \mathbf{U}_2 \rangle$,

(15) $\langle \mathbf{U}_9 \rangle \leq \langle \mathbf{T}_2 \rangle \vee \langle \mathbf{U}_5 \rangle$,

(16) $\langle \mathbf{T}_3 \rangle \leq \langle \mathbf{U}_{10} \rangle$,

(17) $\langle \mathbf{U}_{10} \rangle \leq \langle \mathbf{U}_5 \rangle$,

(18) $\langle \mathbf{U}_{10} \rangle \leq \langle \mathbf{U}_7 \rangle$,

(19) $\langle \mathbf{U}_{10} \rangle \leq \langle \mathbf{U}_9 \rangle$,

(20) $\langle \mathbf{U}_{10} \rangle \leq \langle \mathbf{T}_1 \rangle \vee \langle \mathbf{U}_8 \rangle$,

(21) $\langle \mathbf{U}_{10} \rangle \leq \langle \mathbf{T}_1 \rangle \vee \langle \mathbf{U}_9 \rangle$,

(22) $\langle \mathbf{U}_{10} \rangle \leq \langle \mathbf{T}_2 \rangle \vee \langle \mathbf{U}_3 \rangle$,

(23) $\langle \mathbf{U}_{10} \rangle \leq \langle \mathbf{T}_2 \rangle \vee \langle \mathbf{U}_7 \rangle$,

(24) $\langle \mathbf{U}_{10} \rangle \leq \langle \mathbf{T}_3 \rangle \vee \langle \mathbf{U}_1 \rangle$,

(25) $\langle \mathbf{U}_{10} \rangle \leq \langle \mathbf{T}_3 \rangle \vee \langle \mathbf{U}_3 \rangle$,

(26) $\langle \mathbf{U}_{10} \rangle \leq \langle \mathbf{T}_3 \rangle \vee \langle \mathbf{U}_8 \rangle$,

(27) $\langle \mathbf{U}_3 \rangle \leq \langle \mathbf{U}_{11} \rangle$,

(28) $\langle \mathbf{U}_6 \rangle \leq \langle \mathbf{U}_{11} \rangle$,

(29) $\langle \mathbf{U}_{11} \rangle \leq \langle \mathbf{U}_1 \rangle \vee \langle \mathbf{U}_6 \rangle$,

(30) $\langle \mathbf{U}_{11} \rangle \leq \langle \mathbf{U}_2 \rangle \vee \langle \mathbf{U}_6 \rangle \leq \langle \mathbf{U}_2 \rangle \vee \langle \mathbf{U}_4 \rangle$,

(31) $\langle \mathbf{U}_8 \rangle \leq \langle \mathbf{U}_{12} \rangle$,

(32) $\langle \mathbf{U}_6 \rangle \leq \langle \mathbf{U}_{12} \rangle$,

(33) $\langle \mathbf{U}_{12} \rangle \leq \langle \mathbf{U}_1 \rangle \vee \langle \mathbf{U}_6 \rangle$,

(34) $\langle \mathbf{U}_{12} \rangle \leq \langle \mathbf{U}_4 \rangle \vee \langle \mathbf{U}_6 \rangle \leq \langle \mathbf{U}_2 \rangle \vee \langle \mathbf{U}_4 \rangle$.

The above information completely determines the lattice $L_q(\mathcal{R}_2)$, though it is a bit too large to draw practically. Figure 7.5 shows the order on the join irreducible quasivarieties in $\mathcal{R}_2$.

Incidentally, the quasivariety $\mathcal{R}_2$ provides further examples that the converse of Theorem 2.30 is false, that a quasivariety $\langle \varepsilon_{\mathbf{T},\alpha} \rangle$ with $\mathbf{T}$ quasicritical and $\alpha \succ \Delta$ need not be meet irreducible when it is not in $\kappa(\mathbf{T})$.

PROPOSITION 7.9. *In* $\mathrm{L_q}(\mathcal{R}_2)$, $\langle \varepsilon_{\mathbf{U}_7,By} \rangle = \langle \varepsilon_{\mathbf{T}_1,Bx} \rangle \wedge \langle \varepsilon_{\mathbf{U}_5,By} \rangle$ *and* $\langle \varepsilon_{\mathbf{U}_9,Ay} \rangle = \langle \varepsilon_{\mathbf{T}_2,Ax} \rangle \wedge \langle \varepsilon_{\mathbf{U}_5,Ay} \rangle$.

PROOF. Indeed, it is easy to see that $Ax \to By$ is equivalent to the conjunction of $Ax \to Bx$ and $Ax \,\&\, Bx \to By$. Alternatively, $\langle Ax \to Bx \rangle$ is characterized by the omission of $\mathbf{T}_1$, $\langle Ax \,\&\, Bx \to By \rangle$ by the omission of $\mathbf{U}_2$ and $\mathbf{U}_5$, and $\langle Ax \to By \rangle$ by the omission $\mathbf{T}_1$ and $\mathbf{U}_5$. Since $\mathbf{T}_1 \le \mathbf{U}_2$, Corollary 3.13 yields the meet.

The second part is symmetric. $\qquad\square$

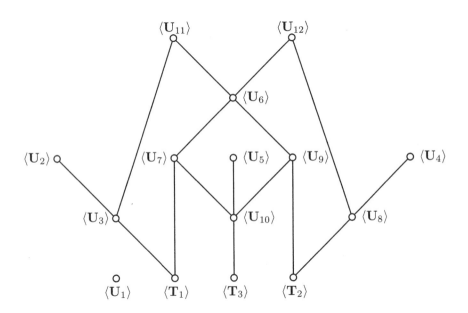

FIGURE 7.5. Join irreducible quasivarieties in $\mathcal{R}_2$ (order only)

With a good deal of patience, the analysis could be extended to quasivarieties $\mathcal{R}_k$ with more predicates, and perhaps patterns found.

7.5. Adding a Constant

An altogether easier option than analyzing $\mathcal{R}_k$ is to add a constant e to the language, with the axioms that $A_j(e)$ holds for all j. If there are k predicates, let us call this quasivariety $\mathcal{E}_k$.

Applying the argument of Lemma 7.2, we see that every quasicritical structure in $\mathcal{E}_k$ has exactly two elements, which we will denote by x and e.

The quasicritical structures in $\mathcal{E}_k$ can be labeled by the subset of the predicates such that Ax holds. For example, in $\mathcal{E}_2$ there are two predicates, say A and B, and the quasicritical structures are $\mathbf{T}_\varnothing$, $\mathbf{T}_A$, $\mathbf{T}_B$, and $\mathbf{T}_{AB}$. Moreover, for distinct subsets X, Y we have $\mathbf{T}_X \not\leq \mathbf{T}_Y$.

The case of $\mathcal{E}_1$ with only one predicate is too simple to illustrate the idea, so let us skip it, noting that there are four quasivarieties: $\mathcal{E}_1$, $\langle Ax \rangle$, $\langle Ax \to x \approx e \rangle$, $\langle x \approx e \rangle$.

The quasivariety $\mathcal{E}_2$ does illustrate the method. The congruence lattice of the free structure $\mathbf{T}_\varnothing$ is given in Figure 7.6. From the congruence lattice we see that $\mathbf{T}_\varnothing \leq \mathbf{T}_A \times \mathbf{T}_B$, which is the only nontrivial subdirect product here. Therefore the lattice of quasivarieties $L_q(\mathcal{E}_2)$ is generated by its four

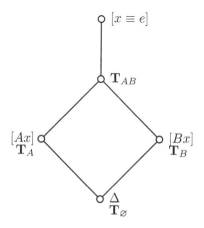

FIGURE 7.6. Congruence lattice of the free 1-generated structure in $\mathcal{E}_2$

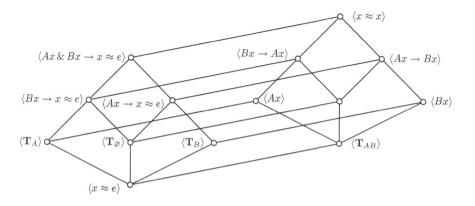

FIGURE 7.7. The lattice of subquasivarieties $L_q(\mathcal{E}_2)$

atoms $\langle \mathbf{T}_\varnothing \rangle$, $\langle \mathbf{T}_A \rangle$, $\langle \mathbf{T}_B \rangle$, $\langle \mathbf{T}_{AB} \rangle$ subject to the relation $\langle \mathbf{T}_\varnothing \rangle \leq \langle \mathbf{T}_A \rangle \vee \langle \mathbf{T}_B \rangle$. The result is the lattice diagrammed in Figure 7.7.

Moreover, it is clear how this works with k predicates. The join irreducible quasivarieties are the atoms $\langle \mathbf{T}_X \rangle$ for $X \subseteq \{A_1, \ldots, A_k\}$. These satisfy the relations that $\langle \mathbf{T}_X \rangle \leq \bigvee_j \langle \mathbf{T}_{Y_j} \rangle$ whenever $\bigcap_j Y_j = X$. We leave it to the reader to describe the meet irreducible quasivarieties, an easy exercise.

CHAPTER 8

Problems

Here are some open questions of varying difficulty and importance.

(1) Can one decide when $\langle \mathbf{T} \rangle$ contains only finitely many subquasivarieties?

Much is known about specific classes of algebras. If $\mathbf{T}$ is a finite quasiprimal algebra, then $L_q(\langle \mathbf{T} \rangle)$ is finite; see Section A.6, also Blanco *et al.* [34]. The atoms of $L_q(\mathcal{K})$ were discussed in Section 4.4. Every 2-element algebra generates an atom, but Adams and Dziobiak [2] exhibit a 3-element algebra that generates a Q-universal quasivariety. Sapir showed that there is a Q-universal quasivariety generated by a finite commutative 3-nilpotent semigroup [152].

The quasivariety generated by a finite abelian group has finitely many subquasivarieties; see Section 2.5. Shakhova determined which nilpotent groups of class 2 generate a quasivariety with only finitely many subquasivarieties [157, 158]. Those nil-2 groups not on her list generate quasivarieties with continuum many subquasivarieties.

Changing the question just slightly, Sapir characterized those locally finite *varieties* $\mathcal{V}$ of groups, rings, and semigroups such that $L_q(\mathcal{V})$ is finite [150, 151]. A locally finite variety $\mathcal{V}$ of groups has only finitely many subquasivarieties if and only if $\mathcal{V}$ is generated by a finite abelian group with abelian Sylow subgroups. A locally finite variety $\mathcal{W}$ of rings has only finitely many subquasivarieties if and only if $\mathcal{W}$ is generated by a finite ring in which every nilpotent subring has zero multiplication. The characterization for semigroups is more complicated: $L_q(\mathbb{V}(\mathbf{S}))$ is finite if and only if $\mathbf{S}$ is one of 20 types of semigroups. For more on varieties generated by small semigroups or monoids, see [61, 88, 89, 113].

(2) Let $\mathbf{T}$ be a finite quasicritical algebra, and let $\mathcal{S}$ be the quasivariety generated by its proper subalgebras. Is there a bound B, perhaps depending on $\mathbf{T}$, such that $\mathcal{S} = \langle \mathbf{T} \rangle_*$ if and only if, for all $n \leq B$,

© Springer International Publishing AG, part of Springer Nature 2018
J. Hyndman, J. B. Nation, *The Lattice of Subquasivarieties of a Locally Finite Quasivariety*, CMS Books in Mathematics,
https://doi.org/10.1007/978-3-319-78235-5_8

$\mathbf{U} \leq \mathbf{T}^n$ implies either $\mathbf{U} \in \mathcal{S}$ or $\mathbf{T} \leq \mathbf{U}$? (Cf. Theorem 2.28, Corollary 2.29, and Theorem 4.12.)

(3) Use the criteria of Theorem 3.10 to prove finite basis results for quasivarieties.

    (a) Does this yield a simple proof of Pigozzi's relatively congruence distributive result [143]? What about the extension to relatively congruence meet-semidistributive quasivarieties by Dziobiak, Maróti, McKenzie, and Nurakunov [60]? Does this extend to relatively congruence pseudo-complemented quasivarieties, perhaps with additional conditions, as in McKenzie and Maróti [120]?

    (b) Can we prove finite basis theorems for finitely generated quasivarieties in a locally finite quasivariety $\mathcal{K}$ in which the relative congruence lattices $\mathrm{Con}_{\mathcal{K}} \mathbf{T}$ are modular? (Cf. [60], [105].)

(4) How can Theorem 3.10 fail? Given a finite algebra $\mathbf{T}$, can there be infinitely many quasivarieties $\langle \mathbf{U} \rangle$ that are minimal with respect to $\langle \mathbf{U} \rangle \nleq \langle \mathbf{T} \rangle$? Sapir has shown that there is a 3-element semigroup that has no upper cover in the lattice of subquasivarieties of the variety it generates [148, 149].

(5) In particular, can $\langle \mathbf{T} \rangle$ have infinitely many upper covers in $\mathrm{L_q}(\mathcal{K})$, when $\mathcal{K}$ is locally finite and has finite type? For such a quasivariety, it is known that $\mathrm{L_q}(\mathcal{K})$ has only finitely many atoms, by Corollary 4.14.

(6) Is it decidable whether a finite algebra in a locally finite quasivariety of finite type generates a finitely based quasivariety? (*No* for varieties by R. McKenzie [118]. See Ježek, Maróti, and McKenzie [90] for related results on quasivarieties.)

(7) Generalizing the interval dismantlable lattices of Section 3.2, a finite lattice is said to be *sublattice dismantlable* if it can be partitioned into two disjoint sublattices, each of which can be partitioned into two disjoint sublattices, etc., until you reach 1-element lattices. Find quasi-identities characterizing this pseudoquasivariety. (Cf. Theorem 3.16.)

(8) Regarding the variety $\mathcal{Z}$ of Section 4.5: Is $\mathcal{Z}$ Q-universal? Does $\langle \mathbf{Z}_1 \rangle$ have an upper cover in $\mathrm{L_q}(\mathcal{Z})$?

(9) Questions relating to the variety $\mathcal{M}$ of Section 5.1:

    (a) Are the quasivarieties $\langle \mathbf{R}_1 \rangle$, $\langle \mathbf{R}_7 \rangle$, $\langle \mathbf{R}_8 \rangle$, or $\langle \mathbf{P}_2 \rangle$ finitely based?

    (b) Find a basis for the finitely based quasivarieties $\langle \mathbf{R}_2 \rangle$ and $\langle \mathbf{R}_4 \rangle$.

    (c) Does $\langle \mathbf{R}_6 \rangle_* = \langle \mathbf{T}_3 \rangle \vee \langle \mathbf{T}_4 \rangle$?

(10) Find better estimates for the maximum number of generators of a quasicritical algebra in $\mathcal{N}_{rs}$. (Cf. Corollary 6.14.)

(11) Can the bound from Theorem 7.5 on the size of a quasicritical pure relational structure be improved?

Properties of Lattices of Subquasivarieties

A.1. Representations of Quasivariety Lattices

The span of nearly two decades since the publication of Viktor Gorbunov's book on quasivarieties [77] has made it desirable to have an updated review of what is known about lattices of quasivarieties. This appendix provides a summary of their basic properties, with references to the literature.

A.2. Representations of Quasivariety Lattices

Representations of a lattice can reveal structural properties, so let us begin there. For a quasivariety $\mathcal{K}$, let $L_q(\mathcal{K})$ denote the lattice of quasivarieties contained in $\mathcal{K}$, ordered by inclusion. No assumption is made yet on the type of the structures in $\mathcal{K}$, but the type will generally have both operations and relations. Of course $L_q(\mathcal{K})$ is a complete lattice.

The *theory* of a quasivariety $\mathcal{K}$ is the set of all quasi-equations holding in $\mathcal{K}$. The theory of a subquasivariety $\mathcal{Q} \leq \mathcal{K}$ contains the theory of $\mathcal{K}$. (More quasi-equations means fewer models.) The lattice $\mathrm{QTh}(\mathcal{K})$ of all quasi-equational theories extending the theory of $\mathcal{K}$, ordered by containment, is dually isomorphic to the lattice of subquasivarieties $L_q(\mathcal{K})$.

In Adaricheva and Nation [19], it is shown that $\mathrm{QTh}(\mathcal{K})$ can be represented as the congruence lattice of a semilattice with operators. That is, there is an algebra $\mathbf{S}$ such that $\mathrm{QTh}(\mathcal{K}) \cong \mathrm{Con}\,\mathbf{S}$, where $\mathbf{S} = (S, \vee, 0, F)$ is a join semilattice with 0 endowed with a set of operations $f : S \to S$ such that $f(x \vee y) = f(x) \vee f(y)$ and $f(0) = 0$ for all $f \in F$. In this representation, $\mathbf{S}$ is the semilattice of compact $\mathcal{K}$-congruences of the free structure $\mathcal{F}_{\mathcal{K}}(\omega)$, and the operators are derived from endomorphisms of the free structure.

In turn, the congruence lattice of a semilattice with operators is dually isomorphic to the lattice $\mathrm{S}_p(\mathbf{L}, H)$ of all H-closed algebraic subsets of an algebraic lattice $\mathbf{L}$, where H is a monoid of operators preserving arbitrary meets and nonempty directed joins (Hyndman, Nation, and Nishida [87]). An *algebraic subset* of a complete lattice is one closed under arbitrary meets and nonempty directed joins. So in the case when $\mathbf{L}$ is finite,

© Springer International Publishing AG, part of Springer Nature 2018
J. Hyndman, J. B. Nation, *The Lattice of Subquasivarieties of a Locally Finite Quasivariety*, CMS Books in Mathematics,
https://doi.org/10.1007/978-3-319-78235-5

$S_p(\mathbf{L}, H)$ is just the subalgebra lattice $\text{Sub}(\mathbf{L}, \wedge, 1, H)$. In the correspondence $\text{Con}(S, \vee, 0, F) \cong^d S_p(\mathbf{L}, H)$, we can take $\mathbf{L}$ to be the filter lattice of $\mathbf{S}$, ordered by set containment, with the operators of H induced by those in F.

Thus, we have two pairs of dual isomorphisms: $L_q(\mathcal{K}) \cong^d \text{QTh}(\mathcal{K})$ and $\text{Con}(S, \vee, 0, F) \cong^d S_p(\mathbf{L}, H)$. Starting with a quasivariety $\mathcal{K}$, we can find a semilattice with operators such that $\text{QTh}(\mathcal{K}) \cong \text{Con } \mathbf{S}$, and hence $L_q(\mathcal{K}) \cong S_p(\mathbf{L}, H)$. Most (but not all) properties of subquasivariety lattices can be derived from this latter representation.

The reverse inclusion is not true: there is a lattice $S_p(\mathbf{L}, H)$ that is not isomorphic to any subquasivariety lattice $L_q(\mathcal{K})$; see the last paragraph of Section A.3. Under certain additional restrictions on the pair $(\mathbf{L}, H)$, for example if $1_{\mathbf{L}}$ is compact and H is a group, one can show that $S_p(\mathbf{L}, H)$ can be represented as $L_q(\mathcal{K})$ for some quasivariety $\mathcal{K}$ [19]. Moreover, if $\mathbf{S}$ is a lattice such that $\mathbf{S} \cong S_p(\mathbf{L}, H)$ for some algebraic lattice $\mathbf{L}$ and monoid of operators H, then the linear sum $\mathbf{1} + \mathbf{S}$ is isomorphic to $L_q(\mathcal{K})$ for a quasivariety $\mathcal{K}$ [17].

Alternatively, if we expand the notion of *quasivariety* to include implicational classes in languages that need not contain equality, with the corresponding adjustments, then every lattice $S_p(\mathbf{L}, H)$ with $\mathbf{L}$ algebraic and H a set of operators is isomorphic to $L_q(\mathcal{I})$ for some implicational class $\mathcal{I}$ [132].

Gorbunov and Tumanov [79] represent a quasivariety lattice $L_q(\mathcal{K})$ as the lattice $S_p(\mathbf{L}, R)$ of all algebraic subsets of an algebraic lattice $\mathbf{L}$ that are closed with respect to a distributive binary relation R, where R corresponds to either isomorphism or embedding of quotients of the lattice of $\mathcal{K}$-congruences of the free structure $\mathcal{F}_{\mathcal{K}}(\omega)$. The representation using operators is perhaps more convenient for describing the structure of $L_q(\mathcal{K})$.

A.3. Basic Consequences of the Representations

The congruence lattice of a semilattice is meet semidistributive and algebraic [142]. So the fact that a subquasivariety lattice $L_q(\mathcal{K})$ is dually isomorphic to the congruence lattice of a semilattice with operators immediately implies that $L_q(\mathcal{K})$ is dually algebraic and join semidistributive, i.e., satisfies

$$(\text{SD}_\vee) \qquad x \vee y \approx x \vee z \to x \vee y \approx x \vee (y \wedge z).$$

Indeed, since being dually algebraic implies lower continuity, $L_q(\mathcal{K})$ satisfies the more general join semidistributive law

$$u \approx x \vee z_i \text{ for all } i \in I \text{ implies } u \approx x \vee \bigwedge_{i \in I} z_i.$$

A complete lattice $\mathbf{K}$ has the *Jónsson-Kiefer Property* if every element $a \in K$ is the join of elements that are (finitely) join prime in the ideal $\downarrow a$. This property holds in all finite join semidistributive lattices [93].

An easy argument shows that in any complete lattice, the Jónsson-Kiefer Property implies join semidistributivity. On the other hand, there is a dually algebraic, join semidistributive lattice that has no join prime elements [18]. Thus the Jónsson-Kiefer Property is strictly stronger than join semidistributivity.

Gorbunov showed that the property holds in lattices of subquasivarieties [76].

COROLLARY A.1. *For any quasivariety* $\mathfrak{Q}$, *the lattice* $L_q(\mathfrak{Q})$ *has the Jónsson-Kiefer Property.*

This is a consequence of a slightly more general result.

THEOREM A.2. *If* $\mathbf{L}$ *is an algebraic lattice and* H *a monoid of operators on* $\mathbf{L}$, *then* $S_p(\mathbf{L}, H)$ *has the Jónsson-Kiefer Property.*

We can sketch the proof of Theorem A.2 as follows. Let $S_p(\mathbf{L})$ denote the lattice of all algebraic subsets of $\mathbf{L}$, with no operators. We use Gorbunov's lemma that the join of finitely many algebraic subsets in the lattice $S_p(\mathbf{L})$ is given by

$$X_1 \vee \cdots \vee X_n = \{x_1 \wedge \cdots \wedge x_n : x_j \in X_j \text{ for } 1 \leq j \leq n\},$$

along with the observation that the join of H-closed algebraic sets is H-closed. Hence if $q \in K$ is meet irreducible, then $q \in X_1 \vee \cdots \vee X_n$ implies $q \in X_j$ for some j. Therefore the H-closed algebraic subset generated by a meet irreducible element q is join prime in $S_p(\mathbf{L}, H)$. But every element of an algebraic lattice is a meet of completely meet irreducible elements. Thus the join of all the join prime algebraic sets in $S_p(\mathbf{L}, H)$ is $\mathbf{L}$ itself. Finally, an algebraic subset of an algebraic lattice is itself an algebraic lattice. Relativizing the argument to a principal ideal $\downarrow A$ in $S_p(\mathbf{L}, H)$ gives the statement in the property.

For a more complete discussion of the Jónsson-Kiefer property, see [18].

Recall that a lattice is *atomic* if it has a least element 0 and for every $x > 0$ there exists an atom a such that $x \geq a \succ 0$.

THEOREM A.3. *For any quasivariety* $\mathfrak{Q}$, *the lattice* $L_q(\mathfrak{Q})$ *is atomic.*

Gorbunov's argument that $L_q(\mathcal{K})$ is atomic lattice [77] uses the special quasivariety $x \approx y$. Let $\mathfrak{T}$ denote the least subquasivariety of $\mathcal{K}$, and let $\mathfrak{T} < \mathfrak{Q} \leq \mathcal{K}$. If it happens that $\mathfrak{Q} \cap \langle x \approx y \rangle = \mathfrak{T}$ (including the case when $\mathfrak{T} = \langle x \approx y \rangle$), then since $\langle x \approx y \rangle$ is finitely based and hence dually compact, there is a quasivariety $\mathcal{H}$ such that $\mathfrak{T} \prec \mathcal{H} \leq \mathfrak{Q}$.

On the other hand, if $\mathfrak{T} < \mathfrak{Q} \cap \langle x \approx y \rangle$, then $\mathfrak{Q}$ contains a 1-element structure $\mathbf{S}$ in which not all the relations of the language of $\mathcal{K}$ hold. Letting $\mathbf{T}_0$ denote the 1-element structure in which all the relations hold, we see that $\{\mathbf{S}, \mathbf{T}_0\}$ is a subquasivariety of $\mathfrak{Q}$ (it is closed under $\mathbb{S}, \mathbb{P}, \mathbb{U}$) which is of course an atom of $L_q(\mathcal{K})$.

We note that a lattice $S_p(\mathbf{L}, H)$ of H-closed algebraic sets need not be atomic. Here is an example from [19]. Let $\mathbf{\Omega}$ be $\omega + 1$ with the single operator h such that $h(k) = k+1$ for $k < \omega$ and $h(\omega) = \omega$. Then $S_p(\mathbf{\Omega}, \{h\}) \cong (\omega+1)^d$, which is not atomic.

A.4. Equaclosure Operators

Wiesław Dziobiak [55] observed that there is a natural closure operator Γ on the lattice of subquasivarieties $L_q(\mathcal{K})$, viz., for $\mathcal{Q} \leq \mathcal{K}$ let $\Gamma(\mathcal{Q}) = \mathcal{K} \cap \mathbb{HSP}(\mathcal{Q})$. In the terminology of Section 4.3, $\Gamma(\mathcal{Q})$ is the least equational subquasivariety of $\mathcal{K}$ containing $\mathcal{Q}$. Moreover, there is a least quasivariety $\mathcal{L} = T(\mathcal{Q})$ with $\Gamma(\mathcal{L}) = \Gamma(\mathcal{Q})$, which is $T(\mathcal{Q}) = \langle \mathcal{F}_\mathcal{Q}(\omega) \rangle$. The map Γ is called the *natural equaclosure operator* on $L_q(\mathcal{K})$, and it has important structural consequences.

We now define an equaclosure operator abstractly to have those properties that we know to hold for the natural equaclosure operator on the lattice of subquasivarieties of a quasivariety. These properties are from Adaricheva and Gorbunov [13], with refinements from Adaricheva and Nation [19], Nation and Nishida [133], and Adaricheva, Hyndman, Nation, and Nishida [17].

An *equaclosure operator* on a dually algebraic lattice $\mathbf{L}$ is a map $\gamma : \mathbf{L} \to \mathbf{L}$ satisfying the following properties. (For the map τ of property (I9), determined by γ, see below.)

(I1) $x \leq \gamma(x)$.
(I2) $x \leq y$ implies $\gamma(x) \leq \gamma(y)$.
(I3) $\gamma^2(x) = \gamma(x)$.
(I4) $\gamma(0) = 0$.
(I5) $\gamma(x) = u$ for all $x \in X$ implies $\gamma(\bigwedge X) = u$.
(I6) $\gamma(x) \wedge (y \vee z) = (\gamma(x) \wedge y) \vee (\gamma(x) \wedge z)$.
(I7) The image $\gamma(L)$ is the complete meet subsemilattice of $\mathbf{L}$ generated by $\gamma(L) \cap C$, where C is the semilattice of dually compact elements of $\mathbf{L}$.
(I8) There is a dually compact element $w \in L$ such that $\gamma(w) = w$ and the interval $[0, w]$ is isomorphic to $S_p(\mathbf{K})$ for some algebraic lattice $\mathbf{K}$. (Thus the interval $[0, w]$ is atomistic.)
(I9) For any index set I, $\gamma(x \wedge \bigvee_{i \in I} \tau(x \vee z_i)) \geq x \wedge \bigvee_{i \in I} \tau(z_i)$.
(I10) If $\tau(x) \leq \tau(y)$ and $\gamma(z) \leq \gamma(y) \leq \gamma(z \vee t)$ and $z \wedge \gamma(x) \leq \gamma(t)$, then $\gamma(x) \leq \gamma(t)$.

Properties (I1)–(I4) say that γ is a closure operator with $\gamma(0) = 0$. Property (I5) means that the operation τ is implicitly defined by γ, *via*

$$\tau(x) = \bigwedge \{z \in L : \gamma(z) = \gamma(x)\}.$$

Thus $\tau(x)$ is the least element z such that $\gamma(z) = \gamma(x)$; more generally, $\gamma(z) = \gamma(x)$ iff $\tau(x) \leq z \leq \gamma(x)$. The element w in (I8) corresponds to the quasivariety $x \approx y$, which for types that include relations need not be the least subquasivariety. When $|I| = 1$, property (I9) becomes simply $\gamma(x \wedge \tau(x \vee z)) \geq x \wedge \tau(z)$, which implies that (I9) holds for all finite index sets.

The corresponding natural operator for a lattice of algebraic subsets $S_p(\mathbf{L}, H)$ can be described easily. Each H-closed algebraic subset A has a least element $a_0 \in L$. Then $\gamma(A) = \uparrow a_0$ is the filter generated by a_0, and $\tau(A)$ is the H-closed algebraic subset generated by a_0.

Property (I8) need not hold for the natural closure operator γ on $S_p(\mathbf{L}, H)$, but the remaining properties (I1)–(I7) and (I9)–(I10) are satisfied. A closure operator satisfying (I1)–(I7) and (I9)–(I10) is called a *weak equaclosure operator*. Moreover, it is these latter properties that give us most (but again not all) the structural consequences.

Not every finite lower bounded lattice admits an equaclosure operator [11]. The two lattices in Figure A.1 are both lower and upper bounded, but support no equaclosure operator. Another example is the lattice $Co(\mathbf{P})$ of convex subsets of the ordered set $\mathbf{P} = \{a, b, c, d, e\}$ with $a < b < c > d > e$. However, every finite lattice which admits an equaclosure operator has a 0-separating homomorphism onto a lower bounded lattice [70].

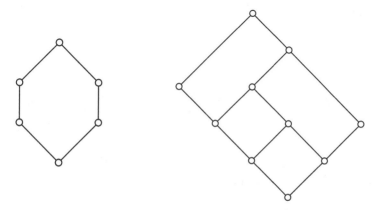

FIGURE A.1. Examples of lower bounded lattices that support no equaclosure operator.

Finally, let us note a property of subquasivariety lattices (from [70], strengthening [76]) that does not require local finiteness, but only that the least element of $L_q(\mathcal{K})$ be dually compact. This will of course be true whenever the type of $\mathcal{K}$ has only finitely many relations.

THEOREM A.4. *If the least element 0 of $L_q(\mathcal{K})$ is dually compact, then $L_q(\mathcal{K})$ satisfies the following quasi-equations for all $n \geq 2$, where $i + 1$ is taken modulo n:*

$$\&_{0 \leq i < n} \, [x_i \leq x_{i+1} \vee y_i \, \& \, x_i \wedge y_i \leq x_{i+1}] \, \& \, x_0 \wedge \cdots \wedge x_{n-1} \approx 0 \rightarrow x_0 \approx 0.$$

The proof uses properties of the equaclosure operator.

A.5. Examples of Quasivariety Lattices

Let us turn to positive results, lattices that *can* be represented as lattices of subquasivarieties. We begin with a classic result of Gorbunov and Tumanov [78].

THEOREM A.5. *The following are equivalent for a complete lattice* **K**.
(1) $\mathbf{K} \cong S_p(\mathbf{L})$ *for some algebraic lattice* **L** *(with no operators).*
(2) $\mathbf{K} \cong L_q(\mathcal{K})$ *for some quasivariety* $\mathcal{K}$ *of one-element relational structures.*

If $\mathbf{K} \cong S_p(\mathbf{L})$ for a finite lattice **L**, then also $\mathbf{K} \cong L_q(\mathcal{R})$ for a quasivariety of rings (Gorbunov [76]).

Recall that a complete lattice is *atomistic* if every element is a join of atoms. Every lattice $S_p(\mathbf{L})$ (with no operators) is atomistic and dually algebraic. The preceding theorem has a nice variation due to Adaricheva, Gorbunov, and Dziobiak [12] that completely characterizes those subquasivariety lattices that are atomistic and algebraic, as well as dually algebraic.

THEOREM A.6. *Let* **K** *be an algebraic, atomistic lattice. The following are equivalent.*
(1) $\mathbf{K} \cong L_q(\mathcal{Q})$ *for some quasi-equational theory* $\mathcal{Q}$.
(2) $\mathbf{K} \cong S_p(\mathbf{L})$ *for some algebraic lattice* **L** *satisfying the ascending chain condition and in which the meet of every infinite descending chain is* 0.
(3) **K** *is a dually algebraic lattice that supports an equaclosure operator.*

By direct construction, Tumanov [165] obtained the following nice result; see also [87].

THEOREM A.7. *Every finite distributive lattice is isomorphic to* $L_q(\mathcal{K})$ *for some quasivariety* $\mathcal{K}$.

It is shown in [17] that every distributive dually algebraic lattice **D** can be represented as $S_p(\mathbf{L}, H)$ with **L** an algebraic lattice and H a monoid of operators. On the other hand, if **D** is not atomic, then it is not isomorphic to $L_q(\mathcal{K})$ for a quasivariety $\mathcal{K}$. The lattice $(\omega + 1)^d$ is one such example.

There are also constructions that yield representations of various particular lattices as $L_q(\mathcal{K})$, e.g., in [17, 19].

A.6. The Quasivariety Generated by a Quasiprimal Algebra

It would be remiss not to discuss a class of very well-behaved algebras. Our summary of primal and quasiprimal algebras follows Bergman [27] (Chapter 6) and Burris and Sankappanavar [39] (Chapter IV). Books by Kaarli and Pixley [95] and Werner [168] cover the closely related topics of functional completeness and discriminator varieties, respectively.

A variety $\mathcal{V}$ is *arithmetical* if it is both congruence-distributive and congruence-permutable. A strong Mal'cev condition due to Pixley characterizes arithmetical varieties [144].

THEOREM A.8. *The following are equivalent for a variety $\mathcal{V}$.*

(1) $\mathcal{V}$ *is arithmetical.*

(2) *There is a term $p(x, y, z)$ such that $\mathcal{V}$ satisfies the identities*

$$p(x, y, x) \approx p(x, y, y) \approx p(y, y, x) \approx x.$$

(3) *There are terms $u(x, y, z)$ and $m(x, y, z)$ such that $\mathcal{V}$ satisfies the identities*

$$u(x, x, y) \approx u(x, y, x) \approx u(y, x, x) \approx x,$$
$$m(x, y, y) \approx m(y, y, x) \approx x.$$

A term $p(x, y, z)$ satisfying the identities of Theorem A.8(2) is called a *Pixley term*. Thus the existence of a Pixley term for a variety (2) is equivalent to the existence of both a majority term and a Mal'cev term (3).

For example, Boolean algebras are arithmetical, where we can take $p(x, y, z) = (x \wedge y') \vee (x \wedge z) \vee (y' \wedge z)$. Michler and Wille showed that a variety of rings is arithmetical if and only if it is generated by a finite number of finite fields [127].

A finite algebra $\mathbf{A}$ is *primal* if, for every $k > 0$, every k-ary function $f : A^k \to A$ can be represented by a term function:

$$f(a_1, \ldots, a_k) = t(a_1, \ldots, a_k)$$

for all $(a_1, \ldots, a_k) \in A^k$. It is a standard exercise (using truth tables) that finite Boolean algebras are primal, and not hard to see (using interpolation) that a field $\mathbf{Z}_p$ of prime order is primal.

The characterization of primal algebras is due to Foster and Pixley [65]. Recall that a structure is *rigid* if it has no non-identity automorphisms.

THEOREM A.9. *A finite algebra $\mathbf{P}$ is primal if and only if $\mathbb{V}(\mathbf{P})$ is arithmetical and $\mathbf{P}$ is simple, rigid, and has no proper subalgebras.*

Note that since a field $\mathbf{Z}_{p^k}$ with $k > 1$ has proper subfields, these are not primal.

Our immediate interest in primal algebras stems from the following result, which is based on a lemma of Fleischer [63], based in turn on Fuchs [72].

LEMMA A.10. *Let $\mathbf{A}$ be an algebra with permuting congruences, and suppose that $\mathbf{A}$ is a subdirect product of finitely many simple algebras. Then $\mathbf{A}$ is isomorphic to a direct product of some of those simple algebras.*

THEOREM A.11. *Let $\mathbf{P}$ be a primal algebra. Then the finite members of $\mathbb{V}(\mathbf{P})$ are (up to isomorphism) precisely the algebras $\mathbf{P}^n$ for $n \in \omega$. Hence $\mathbb{V}(\mathbf{P}) = \mathbb{Q}(\mathbf{V})$, and $\mathrm{L_q}(\mathbb{Q}(\mathbf{P})) = \mathrm{L_v}(\mathbb{V}(\mathbf{P})) \cong \mathbf{2}$.*

Theorem A.11 is due to Foster [64]. The infinite members of $\mathbb{Q}(\mathbf{P})$ may be boolean powers, but since quasivarieties are determined by their finitely generated members and $\mathbb{Q}(\mathbf{P})$ is locally finite, we do not need that description.

The extension to finite sets of primal algebras (with a common Pixley term) is straightforward. This would include, for example, finitely many finite fields of different prime orders.

THEOREM A.12. *Let* $\mathcal{B} = \{\mathbf{P}_1, \dots, \mathbf{P}_n\}$ *be a finite collection of primal algebras contained in a common arithmetical variety. Then the finite members of* $\mathbb{V}(\mathcal{B})$ *are finite direct products of algebras from* $\mathcal{B}$. *Hence* $\mathbb{V}(\mathcal{B}) = \mathbb{Q}(\mathcal{B})$, *and* $\mathrm{L_q}(\mathbb{Q}(\mathcal{B})) = \mathrm{L_v}(\mathbb{V}(\mathcal{B})) \cong \mathbf{2}^n$.

We now turn our attention to quasiprimal algebras, a generalization of primality that retains the choicest properties.

The *ternary discriminator function* $t(x, y, z)$ is an operation defined on any nonempty set by

$$t(x, y, z) = \begin{cases} z & \text{if } x = y, \\ x & \text{otherwise.} \end{cases}$$

Note that the ternary discriminator is particularly simple Pixley function (as in Theorem A.8(2)), and $t(x, y, z)$ is a term function on any primal algebra, since they are functionally complete.

A finite algebra $\mathbf{A}$ is *quasiprimal* if the ternary discriminator is a term function on $\mathbf{A}$. The characterization of quasiprimal algebras is due to Pixley [145].

THEOREM A.13. *A finite algebra* $\mathbf{A}$ *is primal if and only if* $\mathbb{V}(\mathbf{A})$ *is arithmetical and every nontrivial subalgebra of* $\mathbf{A}$ *is simple.*

Every finite field is quasiprimal (though only the fields of prime order are primal). For a field of order n, the discriminator term is

$$t(x, y, z) = (x - y)^{n-1} \cdot (x - z) + z.$$

The extension of Theorem A.12 to quasiprimal algebras is due to Keimel and Werner [108], and Bulman-Fleming and Werner [38].

THEOREM A.14. *Let* $\mathcal{C} = \{\mathbf{A}_1, \dots, \mathbf{A}_n\}$ *be a finite collection of quasiprimal algebras contained in a common arithmetical variety. Then the finite members of* $\mathbb{V}(\mathcal{C})$ *are finite direct products of subalgebras of* $\mathbf{A}_1, \dots, \mathbf{A}_n$. *Hence* $\mathbb{V}(\mathcal{C}) = \mathbb{Q}(\mathcal{C})$, *and* $\mathrm{L_q}(\mathbb{Q}(\mathcal{C})) = \mathrm{L_v}(\mathbb{V}(\mathcal{C}))$ *is a finite distributive lattice.*

Again, the infinite members of $\mathbb{V}(\mathcal{C})$ may be represented as boolean products.

This short survey just scratches the surface of a classical theory. For further study, we recommend the sections on *semisimple varieties* and *directly representable varieties* in the textbooks by Bergman [27] and Burris and Sankappanavar [39].

In 1975, Murskiĭ proved the remarkable fact that almost every finite algebra is quasiprimal [129]. The situation can be sketched as follows. There is a natural probability measure on the space of all finite algebras of a given type; see [27, 68]. Fixing a type τ and a property X, let $\mathrm{Pr}_\tau(X)$ be the probability that a random algebra has the property X. Murskiĭ's theorem can then be stated thusly. Part (1) is an unpublished result of R. O. Davies from 1968.

THEOREM A.15. *Let P be the property of being primal, and Q the property of being quasiprimal.*

  (1) *If the type τ has a single operation symbol of arity $k > 1$, then* $\Pr_\tau(P) = 1/e$.

  (2) *If the type τ has a single operation symbol of arity $k > 1$, then* $\Pr_\tau(Q) = 1$.

  (3) *If the type τ has at least two operation symbols, at least one of which has arity $k > 1$, then* $\Pr_\tau(P) = 1$.

The nice proof in Bergman [27] incorporates unpublished ideas of Quackenbush and McKenzie.

While the quasivariety generated by a quasiprimal algebra has well-behaved subquasivarieties, we have seen that this is not the case for other finitely generated quasivarieties. Congruence distributivity is not enough: the quasivariety generated by a finite lattice can be Q-universal (Theorem 2.41). Congruence permutability is not enough: Nurakunov has shown that the quasivariety generated by a finite abelian group with an extra constant can be Q-universal [137]. Even a quasivariety with a discriminator term is not enough: Dziobiak has exhibited a (non-locally-finite) quasivariety $\mathcal{S}$ generated by a set of quasiprimal algebras such that $L_q(\mathcal{S})$ is uncountable and satisfies no nontrivial lattice identity [53]. It is the combination of these factors that make the quasivariety generated by a finite quasiprimal algebra so nice!

A.7. Miscellaneous

We conclude with references to the literature for some other topics not addressed in this monograph, with no pretense of being comprehensive.

Q-universal quasivarieties were discussed in Section 5.5. In the category theoretical sense, a quasivariety $\mathcal{K}$ is said to be *universal* if every category of algebras of finite type, or equivalently, the category $\mathcal{G}$ of directed graphs, is isomorphic to a full subcategory of $\mathcal{K}$. If the embedding of $\mathcal{G}$ into $\mathcal{K}$ may be effected by a functor which assigns a finite algebraic system to each finite graph, then $\mathcal{K}$ is said to be *finite-to-finite universal*. In [5], Adams and Dziobiak showed that every finite-to-finite universal category is Q-universal. This connection is further explored in Adams and Dziobiak [8].

Relatively congruence distributive quasivarieties were characterized by Nurakunov [135]. The closely related property of a quasivariety having *equationally definable principal congruence meets* was considered by Czelakowski and Dziobiak [46]. Kearnes and McKenzie characterized relatively congruence modular quasivarieties, and developed a commutator theory for these [105].

Shafaat showed that 2-element algebras generate quasivarieties with 2-element subquasivariety lattices, i.e., atoms in $L_q(\mathcal{K})$ [159]. This is of course not true for relational structures: even 1-element relational structures of finite type can generate a quasivariety with a large (finite) subquasivariety lattice. Adams and Dziobiak were the first to show that a 3-element algebra can generate a Q-universal quasivariety [2]; our algebra $\mathbf{T}_3$ of Chapter 5 is only the latest in a long series of such.

For further results on minimal quasivarieties, see Adams and Dziobiak [4], and Bergman and McKenzie [28].

Dziobiak showed that atoms in the lattice of subquasivarieties enjoy special properties [55]. A lattice is *biatomic* if whenever $p \succ 0$ and $p \leq a \vee b$, then there exist atoms $q \leq b$ and $r \leq c$ with $p \leq q \vee r$. Dziobiak's condition was refined by Adaricheva and Gorbunov [13], who showed that lattices of subquasivarieties $L_q(\mathcal{K})$ are biatomic. In fact, any dually algebraic lattice supporting an equaclosure operator is biatomic. This and other restrictions on the atoms, originally from [13], are derived from the property of having a weak equaclosure operator in Adaricheva and Nation [19].

Kearnes and Nation showed that the class of lattices that can be embedded into subquasivariety lattices is not first-order axiomatizable [106]. Schwidefsky has addressed the complexity of quasivariety lattices with the following result [153]. Let $\mathcal{K}$ be a quasivariety of finite type that satisfies a certain generalization of properties (P1)–(P4) for Q-universality from Section 5.5. Then there is a subquasivariety $\mathcal{Q} \leq \mathcal{K}$ such that the problem of embedding a finite lattice into $L_q(\mathcal{Q})$ is undecidable. Related results are in Nurakunov [136] and Schwidefsky and Zamojska-Dzienio *et al.* [25, 155, 156].

There are some very interesting papers on quasivarieties of specific types of algebras, many of which have not come up elsewhere in the text. Here are references for a few.

- Semigroups: Adams and Dziobiak [7], McKenzie [116], Sapir [147, 148, 149, 151, 152].
- Semilattices with operators: Blok and Dziobiak [35], Dziobiak [57], Dziobiak, Ježek, and Maróti [58], Nagy [130].
- MV-algebras: Gispert and Torrens [73], Gispert [74].
- Regularized varieties: Bergman and Romanowska [29].
- Semiprimal varieties: Dziobiak [53].

We can only hope that this monograph has whetted the reader's interest enough to pursue a favorite quasivariety!

Bibliography

[1] M.E. Adams, W. Dziobiak, Q-universal quasivarieties of algebras. Proc. Am. Math. Soc. **120**, 1053–1059 (1994)

[2] M.E. Adams, W. Dziobiak, Lattices of quasivarieties of 3-element algebras. J. Algebra **166**, 181–210 (1994)

[3] M.E. Adams, W. Dziobiak, Quasivarieties of distributive lattices with a quantifier. Discrete Math. **135**, 15–28 (1994)

[4] M.E. Adams, W. Dziobiak, Joins of minimal quasivarieties. Stud. Logica **54**, 371–389 (1995)

[5] M.E. Adams, W. Dziobiak, Finite-to-finite universal quasivarieties are Q-universal. Algebra Univers. **46**, 253–283 (2001)

[6] M.E. Adams, W. Dziobiak, The lattice of quasivarieties of undirected graphs. Algebra Univers. **47**, 7–11 (2002)

[7] M.E. Adams, W. Dziobiak, Quasivarieties of idempotent semigroups. Int. J. Algebra Comput. **13**, 733–752 (2003)

[8] M.E. Adams, W. Dziobiak, Universal quasivarieties of algebras, in *Proceedings of the 9th "Dr. Antonio A. R. Monteiro" Congress, Actas Cong. "Dr. Antonio A. R. Monteiro"* (Univ. Nac. del Sur, Baía Blanca, 2008), pp. 11–21

[9] M.E. Adams, W. Dziobiak, M. Gould, J. Schmid, Quasivarieties of pseudocomplemented semilattices. Fund. Math. **146**, 295–312 (1995)

[10] K. Adaricheva, Characterization of lattices of subsemilattices. Algebra Logic **30**, 385–404 (1991)

[11] K. Adaricheva, W. Dziobiak, V.A. Gorbunov, Finite atomistic lattices that can be represented as lattices of quasivarieties. Fund. Math. **142**, 19–43 (1993)

[12] K. Adaricheva, W. Dziobiak, V.A. Gorbunov, Algebraic point lattices of quasivarieties. Algebra Logic **36**, 213–225 (1997)

[13] K. Adaricheva, V.A. Gorbunov, Equational closure operator and forbidden semidistributive lattices. Siberian Math. J. **30**, 831–849 (1989)

[14] K. Adaricheva, V.A. Gorbunov, On lower bounded lattices. Algebra Univers. **46**, 203–213 (2001)

[15] K. Adaricheva, V.A. Gorbunov, V.I. Tumanov, Join-semidistributive lattices and convex geometries. Adv. Math. **173**, 1–49 (2003)

© Springer International Publishing AG, part of Springer Nature 2018

J. Hyndman, J. B. Nation, *The Lattice of Subquasivarieties of a Locally Finite Quasivariety*, CMS Books in Mathematics, https://doi.org/10.1007/978-3-319-78235-5

[16] K. Adaricheva, J. Hyndman, S. Lempp, J.B. Nation, Interval dismantlable lattices. Order 35, 133–137 (2018)

[17] K. Adaricheva, J. Hyndman, J.B. Nation, J. Nishida, A primer of subquasivariety lattices. Preprint. http://math.hawaii.edu/~jb/

[18] K. Adaricheva, M. Maróti, R. Mckenzie, J.B. Nation, E. Zenk, The Jónsson-Kiefer property. Stud. Logica 83, 111–131 (2006)

[19] K. Adaricheva, J.B. Nation, Lattices of quasi-equational theories as congruence lattices of semilattices with operators, Parts I and II. Int. J. Algebra Comput. 22, N7 (2012)

[20] K. Adaricheva, J.B. Nation, Lattices of algebraic subsets and implicational classes, in *Lattice Theory: Special Topics and Applications*, vol. 2, Chapter 4, ed. by G. Grätzer, F. Wehrung (Brikhäuser, Cham, 2016)

[21] K. Adaricheva, J.B. Nation, R. Rand, Ordered direct implicational basis of a finite closure system. Discrete Appl. Math. 161, 707–723 (2013)

[22] M. Ajtai, On a class of finite lattices. Period. Math. Hung. 4, 217–220 (1973)

[23] C.J. Ash, Pseudovarieties, generalized varieties and similarly described classes. J. Algebra 92, 104–115 (1985)

[24] K. Baker, Finite equational bases for finite algebras in a congruence-distributive equational class. Adv. Math. 24, 207–243 (1977)

[25] A. Basheyeva, A. Nurakunov, M. Schwidefsky, A. Zamojska-Dzienio, Lattices of subclasses, III. Siberian Elektron. Mat. Izv. 14, 252–263 (2017)

[26] V.P. Belkin, Quasi-identities of finite rings and lattices. Algebra Logika 17, 247–259, 357 (1978, in Russian)

[27] C. Bergman, *Universal Algebra: Fundamentals and Selected Topics* (CRC, Boca Raton, 2012)

[28] C. Bergman, R. McKenzie, Minimal varieties and quasivarieties. J. Aust. Math. Soc. (Ser. A) 48, 133–147 (1990)

[29] C. Bergman, A. Romanowska, Subquasivarieties of regularized varieties. Algebra Univers. 36, 536–563 (1996)

[30] I.P. Bestsennyi, Quasi-identities of finite unary algebras. Algebra Logic 28, 327–340 (1989)

[31] G. Birkhoff, On the structure of abstract algebras. Proc. Camb. Philos. Soc. 31, 432–454 (1935)

[32] G. Birkhoff, *Lattice Theory* (American Mathematical Society, Providence, 1940). Revised editions 1948, 1967

[33] G. Birkhoff, Universal algebra, in *Proc. First Canad. Math. Cong.*, Montreal, 1945 (University of Toronto Press, Toronto, 1946), pp. 310–326

[34] J. Blanco, M. Campercholi, D. Vaggione, The subquasivariety lattice of a discriminator variety. Adv. Math. 159, 18–50 (2001)

[35] W. Blok, W. Dziobiak, On the lattice of quasivarieties of Sugihara algebras. Stud. Logica 45, 275–280 (1986)

[36] W. Blok, D. Pigozzi, A finite basis theorem for quasivarieties. Algebra Univers. **22**, 1–13 (1986)

[37] R. Bryant, The laws of finite pointed groups. Bull. Lond. Math. Soc. **14**, 119–123 (1982)

[38] S. Bulman-Fleming, H. Werner, Equational compactness in quasiprimal varieties. Algebra Univers. **7**, 33–46 (1977)

[39] S. Burris, H.P. Sankappanavar, *A Course in Universal Algebra* (Springer, New York, 1981). http://www.math.uwaterloo.ca/~snburris/htdocs/ualg.html

[40] X. Caicedo, Finitely axiomatizable quasivarieties of graphs. Algebra Univers. **34**, 314–321 (1995)

[41] W.H. Carlisle, Some problems in the theory of semigroup varieties. PhD Thesis, Emory University (1970)

[42] D. Casperson, J. Hyndman, Primitive positive formulas preventing a finite basis of quasi-equations. Int. J. Algebra Comput. **19**, 925–935 (2009)

[43] D. Casperson, J. Hyndman, J. Mason, J. Nation, B. Schaan, Existence of finite bases for quasi-equations of unary algebras with 0. Int. J. Algebra Comput. **25**, 927–950 (2015)

[44] C.C. Chang, J. Kiesler, *Model Theory* (North Holland, Amsterdam, 1990)

[45] C.C. Chang, A. Morel, On closure under direct products. J. Symb. Logic **23**, 149–154 (1959)

[46] J. Czelakowski, W. Dziobiak, Congruence distributive quasivarieties whose finitely subdirectly irreducible members form a universal class. Algebra Univers. **27**, 128–149 (1990)

[47] B. Davey, H. Priestley, *Introduction to Lattices and Order* (Cambridge University Press, Cambridge, 2002)

[48] A. Day, Splitting lattices generate all lattices. Algebra Univers. **7**, 163–170 (1977)

[49] A. Day, Characterizations of finite lattices that are bounded-homomorphic images or sublattices of free lattices. Can. J. Math. **31**, 69–78 (1979)

[50] R. Dedekind, *Uber Zerlegungen von Zahlen durch ihre grössten gemeinsamen Teiler* (Festschrift der Herzogl. technische Hochschule zur Naturforscher-Versammlung, Braunschweig, 1897). Reprinted in Gesammelte mathematische Werke, vol. 2 (Chelsea, New York, 1968), pp. 103–148

[51] P. Dellunde, R. Jansana, Some characterization theorems for infinitary universal Horn logic without equality. J. Symb. Logic **61**, 1242–1260 (1996)

[52] R.P. Dilworth, Some combinatorial problems in partially ordered sets, in *Proc. of Symposia in Applied Math.*, vol. 10 (American Mathematical Society, Providence, 1960), pp. 85–90

[53] W. Dziobiak, On subquasivariety lattices of semiprimal varieties. Algebra Univers. **20**, 127–129 (1985)

[54] W. Dziobiak, On lattice identities satisfied in subquasivariety lattices of modular lattices. Algebra Univers. **22**, 205–214 (1986)

[55] W. Dziobiak, On atoms in the lattice of quasivarieties. Algebra Univers. **24**, 32–35 (1987)

[56] W. Dziobiak, Finite bases for finitely generated, relatively congruence distributive quasivarieties. Algebra Univers. **28**, 303–323 (1991)

[57] W. Dziobiak, Quasivarieties of Sugihara semilattices with involution. Algebra Logic **39**, 26–36 (2000)

[58] W. Dziobiak, J. Ježek, M. Maróti, Minimal varieties and quasivarieties of semilattices with one automorphism. Semigroup Forum **78**, 253–261 (2009)

[59] W. Dziobiak, A.V. Kravchenko, P. Wojciechowski, Equivalents for a quasivariety to be generated by a single structure. Stud. Logica **91**, 113–123 (2009)

[60] W. Dziobiak, M. Maróti, R. McKenzie, A. Nurakunov, The weak extension property and finite axiomatizability for quasivarieties. Fund. Math. **202**, 199–223 (2009)

[61] C. Edmunds, Varieties generated by semigroups of order four. Semigroup Forum **21**, 67–81 (1980)

[62] T. Evans, The lattice of semigroup varieties. Semigroup Forum **2**, 1–43 (1971)

[63] I. Fleischer, A note on subdirect products. Acta Math. Acad. Sci. Hung. **6**, 463–465 (1955)

[64] A. Foster, Generalized "Boolean" theory of universal algebras, Part II: Identities and subdirect sums of functionally complete algebras. Math. Z. **59**, 191–199 (1953)

[65] A. Foster, A. Pixley, Semi-categorical algebras II. Math. Z. **85**, 169–184 (1964)

[66] T. Frayne, A. Morel, D. Scott, Reduced direct products. Fund. Math. **51**, 195–228 (1962)

[67] R. Freese, An application of Dilworth's lattice of maximal antichains. Discrete Math. **7**, 107–109 (1974)

[68] R. Freese, On the two kinds of probability in algebra. Algebra Univers. **27**, 70–79 (1990)

[69] R. Freese, J. Ježek, J.B. Nation, *Free Lattices*. Mathematical Surveys and Monographs, vol. 42 (American Mathematical Society, Providence, 1995)

[70] R. Freese, K. Kearnes, J.B. Nation, Congruence lattices of congruence semidistributive algebras, in *Lattice Theory and Its Applications* (Darmstadt, 1991), pp. 63–78; Res. Exp. Math., vol. 23 (Heldermann, Lemgo, 1995)

[71] R. Freese, J.B. Nation, Congruence lattices of semilattices. Pac. J. Math. **49**, 51–58 (1973)

[72] L. Fuchs, On subdirect unions, I. Acta Math. Acad. Sci. Hung. **3**, 103–120 (1952)

[73] J. Gispert, A. Torrens, Locally finite quasivarieties of MV-algebras. arXiv:1405.7504v1 (2014)

[74] J. Gispert, Least V-quasivarieties of MV-algebras. Fuzzy Sets Syst. **292**, 274–284 (2016)

[75] V.A. Gorbunov, The canonical decompositions in complete lattices. Algebra Logic **17**, 323–332 (1978)

[76] V.A. Gorbunov, The structure of lattices of quasivarieties. Algebra Univers. **32**, 493–530 (1994)

[77] V.A. Gorbunov, *Algebraic Theory of Quasivarieties* (Plenum, New York, 1998)

[78] V.A. Gorbunov, V.I. Tumanov, A class of lattices of quasivarieties. Algebra Logic **19**, 38–52 (1980)

[79] V.A. Gorbunov, V.I. Tumanov, *Construction of Lattices of Quasivarieties*. Math. Logic and Theory of Algorithms. Trudy Inst. Math. Sibirsk. Otdel. Adad. Nauk SSSR, vol. 2 (Nauka, Novosibirsk, 1982), pp. 12–44

[80] G. Grätzer, Equational classes of lattices. Duke Math. J. **33**, 613–622 (1966)

[81] G. Grätzer, *Lattice Theory: Foundation* (Springer, New York, 2011)

[82] G. Grätzer, H. Lakser, The lattice of quasivarieties of lattices. Algebra Univers. **9**, 102–115 (1979)

[83] D. Hobby, R. McKenzie, *The Structure of Finite Algebras, Contemporary Mathematics* vol. 76 (American Mathematical Society, Providence, 1988)

[84] A. Horn, On sentences which are true of direct unions of algebras. J. Symb. Logic **16**, 14–21 (1951)

[85] J. Hyndman, Positive primitive formulas preventing enough algebraic operations. Algebra Univers. **52**, 303–312 (2004)

[86] J. Hyndman, R. McKenzie, W. Taylor, k-ary monoids of term operations. Semigroup Forum **44**, 21–52 (1992)

[87] J. Hyndman, J.B. Nation, J. Nishida, Congruence lattices of semilattices with operators. Stud. Logica **104**, 305–316 (2016)

[88] M. Jackson, Finite semigroups whose varieties have uncountably many subvarieties. J. Algebra **228**, 512–535 (2000)

[89] M. Jackson, E.W.H. Lee, Monoid varieties with extreme properties. arXiv:1511.08239v3 (2015)

[90] J. Ježek, M. Maroti, R. McKenzie, Quasiequational theories of flat algebras. Czechoslovak Math. J. **55**, 665-675 (2005)

[91] B. Jónsson, Algebras whose congruence lattices are distributive. Math. Scand. **21**, 110–121 (1967)

[92] B. Jónsson, Equational classes of lattices. Math. Scand. **22**, 187–196 (1968)

[93] B. Jónsson, J. Kiefer, Finite sublattices of a free lattice. Can. J. Math **14**, 487–497 (1962)

[94] B. Jónsson, J.B. Nation, A report on sublattices of a free lattice, in *Contributions to Universal Algebra*. Coll. Math. Soc. János Bolyai, vol. 17 (North-Holland Publishing Co., 1977), pp. 223–257

[95] K. Kaarli, A. Pixley, *Polynomial Completeness in Algebraic Systems* (Chapman & Hall/CRC, Boca Raton, 2001)

[96] V.K. Kartashov, Quasivarieties of unars. Matemasticheskie Zametki **27**, 7–20 (1980, in Russian); English translation in Math. Notes **27** (1980), 5–12

[97] V.K. Kartashov, Quasivarieties of unars with a finite number of cycles. Algebra Logika **19**, 173–193, 250 (1980, in Russian); English translation in Algebra Logic **19**, 106–120 (1980)

[98] V.K. Kartashov, Lattices of quasivarieties of unars. Sibirsk. Mat. Zh. **26**, 49–62, 223 (1985, in Russian); English translation in Siberian Math. J. **26** (1985), 346–357

[99] V.K. Kartashov, Characterization of the lattice of quasivarieties of the algebras $\mathfrak{A}_{1,1}$, in *Algebraic Systems* (Volgograd. Gos. Ped. Inst., Volgograd, 1989, in Russian), pp. 37–45

[100] V.K. Kartashov, On the finite basis property of varieties of commutative unary algebras. Fundam. Prikl. Mat. **14**, 85–89 (2008, in Russian; Russian summary); English translation in J. Math. Sci. (N.Y.) **164**, 56–59 (2010)

[101] V.K. Kartashov, On some results and unsolved problems in the theory of unary algebras. Chebyshevskiĭ Sb. **12**(2), 1–26 (2011). ISBN: 978-5-87954-630-9

[102] V.K. Kartashov, S.P. Makaronov, Quasivarieties of unars with zero, in *Algebraic Systems* (Volgograd. Gos. Ped. Inst., Volgograd, 1989, in Russian), pp. 139–143

[103] K. Kearnes, Relatively congruence modular quasivarieties of modules. arXiv:1509.03809v1 (2015)

[104] K. Kearnes, E. Kiss, *The Shape of Congruence Lattices*. Memoirs of the American Mathematical Society, no. 1046 (American Mathematical Society, Providence, 2013)

[105] K. Kearnes, R. McKenzie, Commutator theory for relatively modular quasivarieties. Trans. Am. Math. Soc. **331**, 465–502 (1992)

[106] K. Kearnes, J.B. Nation, Axiomatizable and nonaxiomatizable congruence prevarieties. Algebra Univers. **59**, 323–335 (2008)

[107] K. Kearnes, Á. Szendrei, A characterization of minimal locally finite varieties. Trans. Am. Math. Soc. **349**, 1749–1768 (1997)

[108] K. Keimel, H. Werner, Stone duality for varieties generated by quasiprimal algebras, in *Recent Advances in the Representation Theory of Rings and C^*-Algebras by Continuous Sections* (Sem., Tulane Univ., New Orleans, 1973). Mem. Amer. Math. Soc., no. 148 (American Mathematical Society, Providence, 1974), pp. 59–85

[109] D. Kelly, I. Rival, Crowns, fences and dismantlable lattices. Can. J. Math. **26**, 1257–1271 (1974)

[110] A.V. Kravchenko, *Q*-universal quasivarieties of graphs. Algebra Logic **41**, 173–181 (2002)

[111] R. Kruse, Identities satisfied by a finite ring. J. Algebra **26**, 298–318 (1973)

[112] J. Lawrence, R. Willard, On finitely based groups and nonfinitely based quasivarieties. J. Algebra **203**, 1–11 (1998)

[113] Y. Luo, W. Zhang, On the variety generated by all semigroups of order three. J. Algebra **334**, 1–30 (2011)

[114] R. Lyndon, Identities in finite algebras. Proc. Am. Math. Soc. **5**, 8–9 (1954)

[115] R. McKenzie, Equational bases and non-modular lattice varieties. Trans. Am. Math. Soc. **174**, 1–43 (1972)

[116] R. McKenzie, The number of nonisomorphic models in quasivarieties of semigroups. Algebra Univers. **16**, 195–203 (1983)

[117] R. McKenzie, Finite equational bases for congruence modular varieties. Algebra Univers. **24**, 224–250 (1987)

[118] R. McKenzie, Tarski's finite basis problem is undecidable. Int. J. Algebra Comput. **6**, 49–104 (1996)

[119] R. McKenzie, G. McNulty, W. Taylor, *Algebras, Lattices and Varieties* (Wadsworth & Brooks/Cole Advanced Books and Software, 1987). Reprinted by AMS Chelsea Publishing, vol. 383, Providence (2018)

[120] M. Maróti, R. McKenzie, Finite basis problems and results for quasivarieties. Stud. Logica **78**, 293–320 (2004)

[121] A.I. Mal'cev, Untersuchungen aus dem Gebiete mathematischen Logik. Rec. Mat. N. S. **1**, 323–335 (1936)

[122] A.I. Mal'cev, Über die Einbettung von assoziativen Systemen in Gruppen, I. Rec. Mat. N. S. **6**, 331–335 (1939)

[123] A.I. Mal'cev, Über die Einbettung von assoziativen Systemen in Gruppen, II. Rec. Mat. N. S. **8**, 251–264 (1940)

[124] A.I. Mal'cev, Several remarks on quasivarieties of algebraic systems. Algebra Logika Sem. **5**, 3–9 (1966, in Russian)

[125] A.I. Mal'cev, Some borderline problems of algebra and logic, in *Proc. Int. Congr. Math.*, Moscow, 1966 (Mir, Moscow, 1968), pp. 217–231

[126] A.I. Mal'cev, *Algebraic Systems (Algebraicheskie sistemy). Sovremennaja Algebra.* Hauptredaktion für physikalisch-mathematische Literatur (Verlag Nauka, Moskau, 1970, in Russian). English translation: Akademie-Verlag, Berlin, 1973. Reprinted by Springer-Verlag, Berlin, 2011

[127] G. Michler, R. Wille, Die primitiven Klasses arithmetischer Ringe. Math. Z. **113**, 369–372 (1970)

[128] V.L. Murskiĭ, The existence in three-valued logic of a closed class without a finite complete system of identities. Doklady Akad. Nauk SSSR **163**, 815–818 (1965)

[129] V.L. Murskiĭ, The existence of finite bases of identities, and other properties of "almost all" finite algebras. Problemy Kibernet **30**, 43–56 (1975, in Russian)

[130] I.V. Nagy, Minimal quasivarieties of semilattices over commutative groups. Algebra Univers. **70**, 309–325 (2013)

[131] J.B. Nation, *Notes on Lattice Theory*. http://math.hawaii.edu/~jb/

[132] J.B. Nation, Lattices of theories in languages without equality. Notre Dame J. Formal Logic **54**, 167–175 (2013)

[133] J.B. Nation, J. Nishida, A refinement of the equaclosure operator. Algebra Univers. **79**, 46 (2018). https://doi.org/10.1007/s00012-018-0518-8

[134] H. Neumann, *Varieties of Groups*. Ergibnisse der Mathematic und ihrer Grenzgebiete, Band 37 (Springer, New York, 1967)

[135] A. Nurakunov, Characterization of relatively congruence distributive quasivarieties of algebras. Algebra Logic **29**, 451–458 (1990)

[136] A. Nurakunov, Unreasonable lattices of quasivarieties. Int. J. Algebra Comput. **22**, 125006, 17 pp (2012)

[137] A. Nurakunov, Lattices of quasivarieties of pointed abelian groups. Algebra Logic **53**, 238–257 (2014)

[138] A. Nurakunov, M. Stronkowski, Quasivarieties with definable relative principal subcongruences. Stud. Logica **92**, 109–120 (2009)

[139] A. Nurakunov, M. Stronkowski, Relation formulas for protoalgebraic equality free quasivarieties: Pałaskińska's theorem revisited. Stud. Logica **101**, 827–847 (2013)

[140] S. Oates, M. Powell, Identical relations in finite groups. J. Algebra **1**, 11-39 (1964)

[141] A.J. Ol'šanskiĭ, Conditional identities in finite groups. Sibirsk. Mat. Zh. **15**, 1409–1413, 1432 (1974, in Russian)

[142] D. Papert, Congruence relations in semilattices. J. Lond. Math. Soc. **39**, 723–729 (1964)

[143] D. Pigozzi, Finite basis theorems for relatively congruence-distributive quasivarieties. Trans. Am. Math. Soc. **310**, 499–533 (1988)

[144] A. Pixley, Distributivity and permutability of congruence relations in equational classes of algebras. Proc. Am. Math. Soc. **14**, 105–109 (1963)

[145] A. Pixley, The ternary discriminator function in universal algebra. Math. Ann. **191**, 167–180 (1971)

[146] P. Pudlák, J. Tůma, Yeast graphs and fermentation of algebraic lattices. Coll. Math. Soc. János Bolyai **14**, 301–341 (1976)

[147] M. Sapir, On the lattice of quasivarieties of idempotent semigroups. Ural. Gos. Univ. Mat. Zap. **11**, 158–169, 213 (1979, in Russian)

[148] M. Sapir, On the quasivarieties generated by finite semigroups. Semigroup Forum **20**, 73–88 (1980)

[149] M. Sapir, Finite and independent axiomatizability of some quasivarieties of semigroups. Soviet Math. (Iz. VUZ) **24**, 83–87 (1980)

[150] M. Sapir, Varieties with a finite number of subquasivarieties. Siberian Math. J. **22**, 934–949 (1981)

[151] M. Sapir, Varieties with a countable number of subquasivarieties. Siberian Math. J. **25**, 148–163 (1984)

[152] M. Sapir, The lattice of quasivarieties of semigroups. Algebra Univers. **21**, 172–180 (1985)

[153] M. Schwidefsky, Complexity of quasivariety lattices. Algebra Logic **54**, 245–257 (2015)

[154] M.V. Semenova, On lattices that are embeddable into lattices of suborders. Algebra Logic **44**, 270–285 (2005)

[155] M. Semenova, A. Zamojska-Dzienio, Lattices of subclasses. Siberian Math. J. **53**, 889–905 (2012)

[156] M. Schwidefsky, A. Zamojska-Dzienio, Lattices of subclasses, II. Int. J. Algebra Comput. **24**, 1099 (2014)

[157] S.A. Shakhova, On the lattice of quasivarieties of nilpotent groups of class 2. Siberian Adv. Math. **7**, 98–125 (1997)

[158] S.A. Shakhova, On the cardinality of the lattice of quasivarieties of nilpotent groups. Algebra Logic **38**, 202–206 (1999)

[159] A. Shafaat, On implicational completeness. Can. J. Math. **26**, 761–768 (1974)

[160] M. Sheremet, Quasivarieties of Cantor algebras. Algebra Univers. **46**, 193–201 (2001)

[161] S.V. Sizyĭ, Quasivarieties of graphs. Siberian Math. J. **35**, 783–794 (1994)

[162] E. Sperner, Ein Satz über Untermengen einer endlichen Menge. Math. Z. **27**, 544–548 (1928)

[163] Á. Szendrei, A survey on strictly simple algebras and minimal varieties, in *Universal Algebra and Quasigroup Theory (Jadwisin, 1989)*. Res. Exp. Math., vol. 19 (Heldermann, Berlin, 1992), pp. 209–239

[164] V.I. Tumanov, Embedding theorems for join-semidistributive lattices, in *Proc. 6th All-Union Conference on Math. Logic*, Tbilisi (1982), p. 188

[165] V.I. Tumanov, Finite distributive lattices of quasivarieties. Algebra Logic **22**, 119–129 (1983)

[166] A.A. Vinogradov, Quasivarieties of abelian groups. Algebra Logika **4**, 15–19 (1965, in Russian)

[167] F. Wehrung, Sublattices of complete lattices with continuity conditions. Algebra Univers. **53**, 149–173 (2005)

[168] H. Werner, *Discriminator Algebras, Studien zur Algebra und ihre Anwendungen*, vol. 6 (Academie-Verlag, Berlin, 1978)

[169] R. Willard, A finite basis theorem for residually finite, congruence meet-semidistributive varieties. J. Symb. Logic **65**, 187–200 (2000)

Symbol Index

© Springer International Publishing AG, part of Springer Nature 2018

J. Hyndman, J. B. Nation, *The Lattice of Subquasivarieties of a Locally Finite Quasivariety*, CMS Books in Mathematics,

https://doi.org/10.1007/978-3-319-78235-5

Author Index

© Springer International Publishing AG, part of Springer Nature 2018   157
J. Hyndman, J. B. Nation, *The Lattice of Subquasivarieties of a Locally Finite Quasivariety*, CMS Books in Mathematics,
https://doi.org/10.1007/978-3-319-78235-5

Subject Index

abelian groups, 32, 34, 44
algebra, 4
algebraic lattice, 2, 9, 13, 34
algebraic subset, 1, 55
  H-closed, 2, 54, 135
arithmetical variety, 140
atomic formula, 4
atomic lattice, 1, 137
atomistic, 140
atoms
  congruences, 21
  subquasivarieties, 57–59, 62

bands, 99
biatomic, 59, 144
branch, 104
branch point, 104

canonical join representation, 2,
  16, 17
Cantor algebras, 99
characteristic congruence, 20, 62
characterized by omitting, 122
closure system, 9, 19
commutative rings, 99
compact, 2, 9, 13, 16, 33, 35
complete lattice, 8
complete meet semilattice, 8
completely join irreducible, 2, 9,
  14, 25

completely join prime, 9, 25, 32,
  39, 98
completely meet irreducible, 2, 9,
  10, 14, 23, 24, 41, 64, 69
completely meet prime, 9, 25
completely simple semigroups, 7
complexity, 144
component, 104
congruence modular, 7, 58
congruence on a relational
  structure, 121
convex subsets, 46, 53, 55, 139
core, 104
critical structure, 34

deMorgan algebras, 99
dependency
  join, 49, 52, 64
  meet, 41, 64
diagram, 20
dimension of a modular lattice,
  26, 27
dismantlable lattice, 47
distributive p-algebras, 99
distributive lattice, 140
distributive lattices with
  quantifier, 99
divisible subset, 44
dual, 8
dual isomorphism, 2, 10, 41, 135
dual refinement, 16, 41

J. Hyndman, J. B. Nation, *The Lattice of Subquasivarieties of a Locally
Finite Quasivariety*, CMS Books in Mathematics,
https://doi.org/10.1007/978-3-319-78235-5

Printed in the United States
By Bookmasters